T0331231

Introduction to Hazard Control Management

A Vital Organizational Function

Introduction to Hazard Control Management

A Vital Organizational Function

JAMES T. TWEEDY

MS, CHCM, CPSM, CHSP

CRC Press
Taylor & Francis Group
Boca Raton   London   New York

CRC Press is an imprint of the
Taylor & Francis Group, an **informa** business

A PRODUCTIVITY PRESS BOOK

CRC Press
Taylor & Francis Group
6000 Broken Sound Parkway NW, Suite 300
Boca Raton, FL 33487-2742

© 2013 by Taylor & Francis Group, LLC
CRC Press is an imprint of Taylor & Francis Group, an informa business

No claim to original U.S. Government works

International Standard Book Number-13: 978-1-4665-5158-9 (Hardback)

Library of Congress Cataloging-in-Publication Data

Tweedy, James T.
  Introduction to hazard control management : a vital organizational function / James T. Tweedy.
    pages cm
  Includes bibliographical references and index.
  ISBN 978-1-4665-5158-9 (hardcover : alk. paper) 1. Industrial safety. I. Title.

T55.T86 2014
658.3'82--dc23
                                                                          2013027850

Visit the Taylor & Francis Web site at
http://www.taylorandfrancis.com

and the CRC Press Web site at
http://www.crcpress.com

This book is dedicated to Harold M. Gordon, founder of the International Board for Certification of Safety Managers (IBFCSM), which is also known as the Board of Certified Hazard Control Management (BCHCM). He served as the board's executive director for more than 30 years and still serves as the chair for the board of directors. His vision about the importance that leadership and management play in controlling hazards and preventing accidents provided the foundation for this text.

Contents

Preface

This introductory text presents a broad overview of hazard control concepts and principles. The text contains information on system safety, understanding organizational cultures, and the importance of leadership. The author presents hazard control as an organizational function and not just a program. Many organizations still refer to their loss-prevention or safety-management efforts as programs. Practicing effective hazard control requires leaders of companies, businesses, and institutions to motivate others to support their loss-prevention efforts. The author did not attempt to present a comprehensive work that addressed all potential hazards and their control. The text does address some common hazards. To prevent and control hazards and prevent harm, leaders, managers, supervisors, hazard control managers, and other key organizational members must make proactive hazard control an organizational priority. The text provides a good foundation for anyone attempting to reduce losses, prevent accidents, and control hazards. The text also addresses the need for good leadership and management. The author also briefly addresses the importance that practicing human-relation and communication skills can have on hazard control efforts. The text also serves as a key reference for those preparing to sit for the Certified Hazard Control Manager Examination. The book uses foundational information first published in the booklet *The Management Approach to Hazard Control*.

James T. Tweedy
International Board for Certification of Safety Managers
Helena, Alabama

Acknowledgments

I would thank Brenda Ammons for her assistance during the early stages of this project. She also created some of the figures and tables found in the text. I would also like to thank Jan Mosier for proofreading each chapter and identifying many of my errors and omissions. Without the help of these dedicated ladies, it would have much taken longer to complete this project. I would also like to express my thanks to Harold Gordon and Harry Stern for their support and encouragement.

Author

James T. Tweedy (Jim) has served as the executive director for the International Board for Certification of Safety Managers (IBFCSM), which is also known as the Board of Certified Hazard Control Management, since 2007. IBFCSM, founded in 1976, offers the following personal credentials: certified healthcare safety professional (CHSP), certified patient safety officer (CPSO), certified hazard control manager (CHCM), certified healthcare emergency professional (CHEP), certified product safety manager (CPSM), and certified hazard control manager-security (CHCM-SEC).

Jim founded TLC Services, a healthcare, hazard-control, organizational-performance, and educational-consulting organization, in 1996. He has more than 25 years of experience with expertise in the areas of credentialing, hazard control, healthcare safety, leadership and team development, education, and compliance. He holds an MS in safety management from Central Missouri University and a BS in liberal studies from Excelsior College. He holds many master-level designations such as CHCM, CPSO, CPSM, CHSP, and CHEP. He also holds a professional membership in the American Society of Safety Engineers and is a member of the American Society of Healthcare Engineering.

James is the author of the best-selling textbook, *Healthcare Hazard Control and Safety Management*, now in its second edition. This book has been used by thousands of healthcare organizations as a daily reference resource. Jim is a polished speaker who presents safety-related topics at seminars, conferences, and in the workplace. He taught more than six years at the college level and is an experienced curriculum developer. He has developed and presented original training programs at locations in more than 40 states. He is a recognized leader in the area of healthcare safety and hazard-control management.

Chapter 1

Hazard Control Concepts and Principles

Introduction

The International Board for Certification of Safety Managers (IBFCSM) defines a *hazard* as "any solid, gas, or liquid with the potential to cause harm when interacting with an array of initiating stimuli including human-related factors." The scope of a hazard can include any activity, behavior, error, event, incident, occurrence, operation, process, situation, substance, or task with potential to cause human harm, property damage, risk to the environment, or a combination of all three. The board defines *hazard closing* as the process of two or more hazards or causal factors attempting to occupy the same space at the same time. Some hazard control professionals refer to this interaction of causal factors as the *accident generation cycle*. IBFCSM (also known as the Board of Certified Hazard Control Management) offers qualified working professionals an opportunity to earn a number of individual certifications including the Certified Hazard Control Manager (CHCM), Certified Product Safety Manager (CPSM), or Certified Healthcare Safety Professional (CHSP) credential. The CHCM credential fills a vital need for a certification appropriate for those working in hazard control–related jobs, positions, and industries. The motto of the CHCM designation, *The Key to Upgrading the Profession*, echoes the importance of using effective leadership concepts and management principles to control hazards in all types of settings.

Accidents, mishaps, and hazardous exposures can result in injuries, illnesses, property damage, and work interruptions. Companies, businesses, and institutions must make hazard control a "priority" organizational function. Proactive hazard control can improve operational efficiency, organizational effectiveness, and the bottom line. The hazard control profession should focus on using management, leadership, and improvement principles to prevent accidents, injuries, and other losses. Senior leaders must ensure that organizational members promptly report accidents, hazards, close call incidents, and unsafe behaviors. Organizations can unknowingly promote activities that do little to improve safety-related behaviors or encourage continuous learning processes. Passive hazard control efforts can communicate a general awareness about the importance of working safely. However, many well-intended initiatives do not achieve measurable results when the organization fails to make hazard control a priority function.

Most organizations must comply with some type of safety and environmental standards. However, making compliance the centerpiece of hazard control efforts can send the wrong message to many organizational members. Once a maintenance supervisor leaving a safety responsibility presentation that I conducted made this comment to me: "I will never again tell any of my technicians to work safely because of compliance, accreditation, or organizational requirements." He then said, "I will tell my subordinates that we will work safely on every task because it is the right thing to do." He decided to become a leader instead of using compliance standards and organizational policies as the key motivators for working safely. When organizational leaders and supervisors make people the priority, adherence to established policies and compliance standards becomes easier to achieve. Most experienced hazard control managers understand the importance that engineering principles play in preventing accidents and injuries. Some well-known engineering innovations such as fire prevention technologies and safer machine designs make workplaces much safer for everyone. Effective hazard control managers must use leadership to minimize risky and unsafe behaviors. Failing to do so can impact morale, operational productivity, and result in higher accident rates. Finally, hazard control must consider using system safety methods. Using a system approach to safety performance considers the need for hazard controls during the design phase of a product, equipment, or process.

Hazard Control Management

This text uses the term *hazard control* instead of the more traditional term of *safety* to describe actions necessary to prevent accidents or reduce losses. Using the phrase *hazard control management* does not diminish the importance of safety and other disciplines such as risk management, occupational health, or industrial hygiene. *Hazard control–related disciplines* can include system safety, transportation safety, consumer safety, and school safety, to name just a few. Hazard control management must focus on developing processes or systems that can help prevent harm and loss. An uncorrected hazard or hazardous situation could contribute to an event resulting in property damage, job interruption, personal harm, or adverse health effects. The process of controlling hazards may require development of written policies, plans, or procedures. However, never consider these written documents as the hazard control management program. Never consider hazard control as a program but as a function of the organization. The hazard control function must connect with organizational structures and operational philosophies including values (Table 1.1).

Table 1.1 Seven Values of Hazard Control Management

- Endless process
- People focused
- Leadership driven
- Operational priority
- Benefits everyone
- Reduces losses
- Prevents human harm

Program or Function

I once asked a safety coordinator to let me review the organizational hazard control plan. He handed me a three-ring binder that came from a bookshelf near his desk. The binder, labeled safety program, contained a number of documents still encased in their original shrink wrap. Program comes from the French word "programme," which means agenda or public notice. We can also refer to the Greek word "graphein" which means to write. When used with the prefix "pro," it became "prographein," which means to write before. Many organizations develop written safety programs to satisfy organizational mandates or to demonstrate visual compliance with regulatory requirements. Written plans, policies, and procedures should direct the hazard control function. The word function, first used in the early sixteenth century, denotes the concept of performance or execution. A function can relate to people, things, and institutions. A function can refer to serving a designated or defined role in some manner. A function can also relate to participation in an ongoing cultural or social system. Considering hazard control as a function of the organization elevates its priority in the minds of everyone (Tables 1.2 and 1.3).

Hazard Control Is Good Business

Liberty Mutual, in its *2007 Workplace Safety Index*, estimated that in 2005 employers paid almost $1 billion per week in direct compensation costs for disabling workplace injuries and illnesses. Senior leaders must make hazard control a priority function. Proactive efforts can help reduce workers' compensation premiums, injury costs, and lost productivity. Liberty Mutual sent a

Table 1.2 Proactive versus Reactive Hazard Control

Proactive	Reactive
Anticipates, recognizes, and identifies hazards	Evaluates and investigates past incidents or accidents
Analyzes and determines risks	Uses risk management to control losses
Controls hazards to reduce accident potential	Satisfied with reducing accident recurrence
Educates and encourages safe behaviors	Disciplines unsafe actions and behaviors
Focuses on preventing losses	Accepts some losses if not too severe
Analyzes to determine root causes	Documents errors and primary causes
Operates to open and hidden cultures	Responsive to formal culture expectations
Leaders involved in hazard control	Leaders delegate responsibilities to others

Table 1.3 Traditional Hazard Control Assumptions

- Hazard control manager retains responsibilities for solving all safety-related problems
- Senior leaders view hazard control as a necessary expense
- Training and education focuses on documentation and not human performance
- Organizational efforts focus on hazards with minimum emphasis on unsafe behaviors

survey to hundreds of chief financial officers in 2005. More than 60% of those responding to the survey indicated that they could document a return on investment for money allocated to hazard control–related initiatives. Occupational Safety and Health Administration (OSHA) reports that the average worksite participating in the OSHA Voluntary Protection Programs documented Days Away, Restricted, or Transferred rates of 52% below the national average for their industrial classification. Organizational leaders making hazard control part of a good business initiative must understand accidents impact their organization in the terms of cost, time, performance, and morale. Proactive hazard control can also help achieve compliance with the myriad of regulatory requirements placed on businesses today.

Hazard Control Responsibilities

Many organizations with high accident or injury rates fail to outline specific hazard control responsibilities in their plans, procedures, directives, and job descriptions. The concept of responsibility relates to a person's obligation to carry out assigned duties in an efficient, effective, and safe manner. Senior leaders must ensure that managers and supervisors understand the importance of their assigned hazard control responsibilities. Senior leaders must ensure that job descriptions address hazard control responsibilities inherent with each position or task. Hazard control efforts will yield results when leaders encourage participation and hold key managers accountable. Dan Petersen, in his second edition of *Safety Management*, stated, "If we study any mass data, we can readily see that the types of accidents resulting in temporary total disabilities are different from the types of accidents resulting in permanent partial disabilities or in permanent total disabilities or fatalities. The causes are different." Professor Peterson makes a very important point. Senior leaders and hazard control managers must learn to focus on the hazards, behaviors, and risks that pose the most potential harm (Table 1.4).

Hazard Control Manager Responsibilities

An effective hazard control manager serves as a consultant and adviser to managers at all operational levels. Hazard control managers must persuade management action, rather than attempt to correct every hazardous situation. The need for improving hazard control efforts must remain proportional to the need for improving other organizational functions. Hazard control objectives must focus on accident prevention, reducing operating costs, and efficiently using human and other organizational resources. Hazard control managers learn to compile and disseminate important safety-related information to managers throughout the organization.

Hazard control managers must teach others about accident-prevention principles and solicit their input. When seeking senior leader approval for hazard control expenditures, use a

Table 1.4 Senior Management Responsibilities

• Develop, sign, and publish an organizational hazard control policy statement
• Describe key expectations related to accomplishing hazard control objectives
• Ensure that all organizational members can explain the major objectives
• Develop methods to track progress and provide feedback to all organizational members
• Require managers and supervisors to visibly support established objectives

Table 1.5   Hazard Control Manager Responsibilities

- Guide development of hazard control training and educational sessions
- Serve as the hazard control consultant and information center
- Provide hazard control-related technical assistance as necessary
- Provide information about legal and compliance requirements affecting safety and health
- Evaluate overall hazard control performance as related to established objectives or goals
- Maintain communication with regulatory agencies and professional safety organizations
- Oversee accident investigations, hazard analysis, and preparation of reports or summaries
- Monitor progress of corrective actions required to address hazards or other safety deficiencies

well-prepared cost-benefit analysis document. Hazard control managers should anticipate opposition from certain segments within their organizations. When dealing with opposition, use effective human relation and communication skills to persuade others to support hazard control objectives.

Hazard control managers should "know what they know" and acknowledge the things they don't know. However, they must know where to go to find answers. Hazard control managers must acknowledge that many operational managers and supervisors face issues beyond their control. Understanding this important concept can help hazard control managers gain their respect. Conducting periodic perception surveys can reveal what people in the organization truly think or believe about hazard control efforts (Table 1.5).

Supervisor Responsibilities

Supervisors must possess the knowledge and experience to provide hazard control guidance to those they lead. First-line supervisors occupy a key hazard control position in many organizations. This position of trust can require supervisors to conduct area inspections, provide job training, ensure timely incident reporting, and accomplish initial accident investigations.

Supervisors in many organizations possess little control over factors such as hiring practices, working conditions, and equipment provided to them. Supervisor must understand the role that human factors can play in accident prevention and causation. They must ensure that each person they supervise understands the behavior expectations of the job. Some organizations require employees to sign a safe work agreement. Such an agreement requires the individual to commit to working safely and adhere to organizational policies or procedures. Supervisor must ensure that their subordinates can access all hazard control plans, policies, and procedures (Table 1.6).

Addressing Behaviors

Supervisors must explain work rules and behavioral expectations to all new or transferred employees. Supervisors must never tolerate individuals who encourage others to disregard work rules or established procedures. When disciplining an individual, do so in private but always document the facts. Senior leaders, managers, and supervisors must set an example for others. They must discourage poor behaviors by reinforcing the importance of acceptable behaviors.

Table 1.6   Supervisor Responsibilities

- Enforce work rules and correct unsafe or at-risk behaviors
- Implement hazard control policies, procedures, and practices in their areas
- Provide job- or task-related training and education
- Immediately report and investigate all accidents in their work areas
- Conduct periodic area hazard control and safety inspections
- Ensure proper maintenance and servicing of all equipment and tools
- Lead by examples and personally adhere to hazard control requirements
- Conduct safety and hazard control meetings on a regular basis
- Work with organizational hazard control personnel to correct and control hazards
- Ensure all personnel correctly use required personal protective equipment

Table 1.7   Behavior Correction Process

- Step 1—Identify the unsafe action
- Step 2—State concern for worker safety
- Step 3—Demonstrate the correct and safe way
- Step 4—Ensure the worker understands
- Step 5—Restate concern for personal safety
- Step 6—Follow-up

Never confuse correcting a behavior with undertaking needed disciplinary action. When correcting an unsafe behavior, always state the facts about the situation but limit personal opinions. Use statements that begin with "I" but never use "they" statements. Take time to recognize good behaviors by using positive reinforcement. Keep in mind that some individuals may not recognize a hazard or hazardous situation. Some may recognize a hazard but not possess the ability to deal with it. Too many injuries occur when a person recognized the hazard but failed to respect its potential for causing harm. Some individuals, for unknown reasons, purposely decide to engage in unsafe or risky behaviors (Table 1.7).

Employee Engagement

Employee engagement occurs when individuals personally feel their connection to their position or job. This engagement also refers to their personal commitment to the success of the organization. Employee engagement can contribute to individual satisfaction and personal mental wellness. Engaged employees also help improve the productivity, morale, and motivation of others. Today, many organizations realize the need for balancing work demands with a person's family and other life issues. Organizational members when off the job serve in a variety of roles including volunteer, caregiver, and parent. Understanding employee engagement helps leaders and hazard control managers deal with the complexity of human behaviors. Conflicting responsibilities can lead to role misunderstandings and work-related overloads that can impact organizational objectives including hazard control efforts.

Working with Employee Organizations

Organizational leaders and hazard control personnel must cooperate with labor groups. When working with labor groups, seek to find common ground for agreement but never sacrifice the principles of hazard control. However, never let hazard control or safety become a bargaining chip during negotiations. Collective bargaining organizations can help by promoting positive worker attitudes. When negotiating, ensure that all safety demands relate to objective criteria. Hazard control personnel must advise management on labor-related hazard control and safety issues. Hazard control personnel should also attempt to participate in collective bargaining sessions when hazard control concerns remain the table.

Hazard Control Practice

Hazard Control Policy Statements

A hazard control policy statement should clearly address an organization's philosophy and objectives related to accident prevention. The policy statement should cover broad hazard control expectations and outline some key responsibilities. The policy statement, when written in precise and unambiguous language, should communicate organizational commitment to a safe and healthy work environment. Senior leadership must sign and disseminate the policy statement to all organizational members. Some organizations publish a well-written policy statement that conflicts with actual operational reality. Ensure the published policy reflects the *real* organizational beliefs and expectations. For most organizations, policy statements should facilitate the decentralization of the hazard control function. The decentralization of hazard control occurs when an organization promotes accident prevention efforts as part of everyone's job. The hazard control manager's role becomes primarily focused on coordinating, promoting, and communicating accident-prevention techniques.

Hazard Control Plan

I believe that many hazard control managers can take some planning tips from emergency and disaster planners. Emergency management planners develop their emergency operations plans by using results obtained from a hazard vulnerability analysis (HVA). Hazard control managers should use a similar approach when beginning to develop their master hazard control directive. Conducting a thorough "hazard vulnerability assessment" would provide a solid foundation on which to build necessary procedures, policies, and action plans. Hazard control plans must "direct" some type of action, intervention, or behavior. Many well-meaning safety programs look good on paper but fail to provide direction on how to reduce hazards, accidents, and injuries. Developing a master hazard control plan based on accurate assessments can provide direction to all accident prevention efforts. The plan should focus on the immediate correction of hazards discovered by the use of hazard assessment data, periodic inspections results, and accident investigation reports.

A master hazard control plan should function as "the directive" for all organizational accident prevention efforts. An effectively written document should provide the roadmap for meeting organizational hazard control expectations, objectives, and goals. A comprehensive plan must hold managers and supervisors accountable for ensuring all operational policies, procedures, and job practices meet the requirements outlined in the master directive. I once discovered, during an on-site hazard control assistance visit, the existence of two "conflicting" OSHA emergency action plans for same facility. During an actual emergency situation, the existence of two plans could contribute to confusion.

Organizations should supplement hazard control–related education with specific on-the-job training. Hazard control managers must monitor a plan's effectiveness. Hazard control plans should stress the importance of establishing procedures for the immediate reporting of accidents, incidents, mishaps, and other "close call" events. Promoting personal ownership of the hazard control function helps promote its importance as an organizational priority. Senior leaders must ensure that key managers provide visible hazard control–function involvement. Poor management practices can contribute directly and indirectly to the generation of accidents. The best written plans will fail leaders tolerate or ignore known management deficiencies or inefficiencies.

I once asked a hazard control coordinator during a consultation visit a four-part question: "What's your most flammable substance, how much do you store on-site, how and where do you store it, and which departments use the substance?" He could only answer part one of the question. The hazard control coordinator did not know or understand that this flammable substance created "fire load" hazard at several locations throughout the facility. This situation created other hazard control issues such as human exposure risks, sprinkler system coverage, proper storage room configuration, and portable fire extinguisher assessments. Once during an on-site survey, I entered a supply room located between two nursing facility "resident" rooms. I discovered more than 100 gallons of isopropyl alcohol (90%) in single gallon containers stacked neatly on wooden shelves. No one could or "would" provide an explanation for this major flammability hazard. Once, I walked into an area of a large facility where noise levels exceeded 100 decibels. My escort never offered me noise protection before we entered the area. The organizations failed to recognize or simply ignored the existence of these hazards. Hazard control personnel, managers, and supervisors must know about the hazards existing under their watch. Hazard control practice not only requires a good action plan but also requires responsible individuals to take responsibility for doing the right thing. Plans, programs, and policies by themselves can't identify and correct hazards or hazardous situations—people do (Table 1.8).

Table 1.8   Content Suggestions for Written Hazard Control Plans

• Outline key duties and responsibilities of managers, supervisors, and organizational members.
• Develop measurement tools and maintenance requirements for all accident prevention efforts.
• Identify the processes used to identify, analyze, and control hazards and unsafe operations.
• Create effective communication and feedback processes to keep everyone informed.
• Establish sound accident investigation methods, procedures, and priorities.
• Ensure the effectiveness of information reporting and collection processes.
• Describe the methods and procedures for hazard analysis and hazard control implementation.
• Provide clear policies about the requirement to adhere to safe work practices and requirements.
• Implement orientation, training, and educational objectives that address "real-world" topics.

Objectives and Goals

Organizations and hazard control managers must ensure the development of realistic objectives and goals. The attainment of hazard control objectives and goals will require development of written plans, procedures, policies, and directives. Develop and implement written documents that "direct" or "require" specific hazard control–related actions and behaviors. Written documents can assign hazard control responsibilities, communicate hazard control issues, and address issues such as inspections, training, job-related processes.

Reviewing Plans, Policies, and Procedures

Conduct periodic reviews of hazard control directives, plans, policies, procedures, and practices. The evaluations must determine reasons that the organizational objectives and goals went unmet. The review should determine if managers, supervisors, and other organizational members actually supported hazard control efforts. Some OSHA standards can require periodic review of written plans and policies. For example, OSHA requires an annual review of the Bloodborne Pathogen Local Exposure Control Plan. When reviewing written plans, place a strong emphasis on accuracy, duplicity, currency, and effectiveness. I suggest that organizations develop a suspense and documentation process to help manage a thorough review of each written hazard control document. Establish a suspense process that includes the information listed as follows for all written plans, policies, and procedures:

- Title of plan, policy, directive, or procedure to include the purpose and date of document.
- Function, division, or office responsible for the document and frequency of review.
- Determine the type of documentation procedures needed.
- Validate standard, directive, or regulation that mandates the document.
- Review the application and scope of the plan (departments, functions, or units).
- Assess the need for special coordination requirements (departments, functions, or units).
- Evaluate all processes used to communicate changes and revisions.

Make minor changes to hazard control plans using ink and inform all interested parties of the change. Develop a log or spreadsheet to maintain a catalog of suggestions for possible future changes to the plan, policy, or procedure. Documenting ink changes and maintaining a log of possible changes can help during the actual revision of the plan. Word processing permits quick updates of plans, policies, and procedures. However, ensure that the written hazard control documents don't become "sanitized versions" for show or display only. Organizations that do not refer to actual hazard control policies, procedures, and plans on an ongoing basis may not need the documents at all. After reviewing hundreds of written hazard control documents during past 20 years, I came to a simple conclusion—most documents experienced little use.

Hazards

Classifying and defining hazards can vary greatly depending on a number of factors, including type of industry, process, or operation. For example, mechanical energy hazards can involve components that cut, crush, bend, shear, pinch, wrap, pull, and puncture. Biological hazards can include pandemic, bioterrorism agents, blood-borne pathogens, and infectious waste. Chemical hazards include substances such as solvents, flammable liquids, compressed gases, cleaning agents, and even disinfectants. Physical hazards are things such as fire, radiation, machine operation, and noise. Environmental and ergonomic hazards include slip, trip, and fall hazards; walking and working surfaces; lighting; and tasks with repetitive motions. Psychosocial hazards address issues such as workplace violence, work-related stress, sleep deprivation, mental problems, chemical dependency, alcohol abuse, and horseplay on the job.

Hazard Identification

Hazard identification requires the identification of hazards, unsafe conditions, and risky behaviors. Hazard anticipation relies on human intuition, training, common sense, observation, and continuous awareness. To identify hazards, rely on the use of inspections, surveys, analysis, and human recognition reporting. Hazard-identification efforts should focus on unsafe conditions, hazards, broken equipment, and human deviations from accepted practices. Require supervisors or unit safety coordinators to conduct periodic area inspections. These individuals should understand hazardous areas and the workers better than anyone. However, supervisors can fall prey to "inspection" bias, which results in poor survey results. Many supervisors conduct limited "ongoing" inspections as part of their daily job duties. Periodic inspections and surveys can focus on critical components of equipment, processes, or systems with a known potential for causing serious injury or illness. Some equipment inspections help meet preventive maintenance requirements or hazard control plan objectives. Safety standards can mandate that "qualified persons" periodically inspect some types of equipment, such as elevators, boilers, pressure vessels, and fire extinguishers, at regular intervals. Establish the frequency of inspections by considering the scope and type of the hazardous operations. Many hazard control plans fail to provide sufficient guidance about how to conduct hazard surveys, inspections, and audits. For example, I know of very few organizations that provide training or education about the proper use of "general" or "demand–response" checklists. Inspections, audits, and hazard surveys can only help identify hazards when conducted properly.

Providing a checklist to an untrained person can result in his or her failure to properly identify hazards or unsafe conditions. General checklists serve as tools that guide an inspection process. These documents do not contain information about all potential hazards. The effective use of "demand–response checklists" will also require some type of education or training. Demand–response checklists address specific operations and complex job processes, such as the operation of robotic systems or the locking out hazardous energy.

Managing Hazards

Organizations covered by OSHA standards must conduct a formal hazard assessment to determine the need for personal protective equipment (PPE). This requirement provides an opportunity for organizations to establish a baseline hazard inventory and assessment. OSHA requires the use of PPE when other controls prove inadequate. A senior manager must certify the accomplishment of the PPE assessment. Hazard control managers can use the information from the PPE hazard assessment to create a *facility hazard index (FHI)*. The FHI should catalog biological, chemical, physical, ergo-environmental, and psychosocial hazards. These categories could be subdivided as needed to provide a better management tool and better hazard documentation. For example, physical hazards could be subdivided into fire hazards, radiation exposure risks, and hearing conservation. Some personnel may wish to structure the FHI using other categories such as safety, security, hazardous materials, patient safety, and occupational safety. The FHI should provide a clear overview of key hazards. Update the list at least annually. Recommend documenting the following information: (1) comprehensive hazard description, (2) date listed, (3) reason listed,

(4) applicable hazard analysis information, (5) controls used to reduce risk, (6) PPE issued to employees, and (7) training or education provided to deal with a hazard.

Controlling hazards that contribute to accidents requires the use of appropriate administrative, work practice, and engineering controls. Encourage and enforce the use of proper PPE when mandated by the job or task. Teach supervisors how to observe, recognize, and correct unsafe behaviors. Maintain all equipment and machinery in top condition. Install and maintain appropriate safety devices or guards. Other important considerations should include designing safe work areas and properly placing machinery to meet human capabilities or limitations. Develop and implement an effective emergency action plan to address injury response, evacuations, and other contingencies.

Organizations should conduct comprehensive baseline and periodic surveys to identify all safety, health, and ergonomic hazards. The results of well-conducted surveys can provide valuable information for use by the hazard control and training functions. Information technology permits organizations to provide timely hazard-related information to its members. Information accessed can include regulations, codes, standards, best industry practices, training material, and hazardous material information. Many manufacturers develop and disseminate hazard information for use by those who buy and operate their equipment and products. The information can include safety warnings, hazard label information, and operational instructions (Tables 1.9 and 1.10).

Preparing for Inspections

Conduct education and training sessions about how to conduct inspections. Periodic inspections provide opportunities for hazard control personnel, line supervisors, and top managers to listen to concerns of those doing the work. Inspections must accurately assess all environments,

Table 1.9   Occupational Hazard Categories

• Biological hazards include bacteria, viruses, infectious waste, and blood-borne pathogens. • Chemical hazards can pose a variety of risks due to their physical, chemical, and toxic properties. • Ergonomic and environmental hazards include repetitive motion, standing, lifting, trips, and falls. • Physical hazards include things such as radiation-, noise-, and machine-generated hazards. • Psychosocial hazards include substance abuse, work-related stress, and workplace violence.

Note: Some hazards may fit in more than one category.

Table 1.10   Common Factors Inherent in Good Work Environments

• Good workplace design and proper equipment placement including guards and controls. • Equipment inspections and preventive maintenance conducted as scheduled. • Organization conducts inspections, audits, and hazard surveys on regular basis. • Corrective actions and hazard controls implemented immediately to eliminate risks. • Employees formally commit to work safely and maintain hazard-free work areas. • Work areas equipped with proper lighting, ventilation, and environmental controls. • Employees must use personal protective equipment when mandated. • Supervisors conduct job instruction, inspections, and initial accident investigations.

equipment, and processes. Don't "just look" for hazards, but learn to observe individuals accomplish specific job tasks or processes. Learning to identify hazards and recognize unsafe behaviors requires inspectors to use their observation skills. Inspectors must focus on using use all five human senses. Look for deviations from accepted work practices and rely on intuition or "gut" feelings to assist with the identification of hazards. Curiosity can help uncover "hidden" hazards. Learning to use visualization techniques to "connect the dots" can create a "mind picture" of a hazardous situation. Never allow human emotions or personal issues to drive the inspection process. Inspectors should maintain a professional demeanor and rely on logic when assessing tough situations. Inspectors must always point out potential or immediate dangers. Inspectors must never operate any equipment unless trained and authorized to do so. Inspectors should ask questions about tasks or processes but refrain from disrupting operations or creating distractions. Well-designed inspection checklists can assist with the documentation of key findings (Table 1.11).

Inspection Reports

When writing an inspection report, provide a summary of key unsafe conditions, unsafe behaviors, and hazards. Ensure the report includes recommendations for correcting all identified hazards. Document hazard locations by department, function, or cost center. Number or title each finding and use a hazard classification description or priority system as mandated by the organization. If possible, assign a priority to each finding. Communicate findings, to appropriate levels of management, in a concise and factual manner. Senior leaders should ensure all actions items receive immediate corrective attention. If a permanent control will take time to implement, ensure that interim controls protect exposed personnel. If possible, specify the recommended corrective action for each identified hazard or unsafe condition (Table 1.12).

Table 1.11   Common Inspection Observations

- Operating vehicle at unsafe speeds or violating safe practice rules
- Removing machine or equipment guards and tampering with safety devices
- Using defective tools and equipment or using them in unsafe ways
- Handling materials in unsafe or careless ways and lifting improperly
- Repairing/adjusting equipment while in motion, under pressure, or electrically charged
- Failing to properly use personal protective equipment
- Unsafe, unsanitary, or unhealthy conditions, including poor housekeeping practices
- Standing or working under suspended loads, scaffolds, shafts, or open hatches

Table 1.12   Categorizing Hazard Correction Priorities

- Category A finding—Major hazards that require immediate correction
- Category B finding—Serious hazards that require short-term correction
- Category C finding—Minor hazard corrected as soon as possible
- Category D finding—Hazard identified but corrected on the spot

Hazard Analysis

Organizations can use a variety of processes to analyze workplace hazards and accident causal factors. Hazard evaluations and accident trend analysis can help improve the effectiveness of established hazard controls. Routine analysis enables an organization to develop and implement appropriate controls for hazardous processes or unsafe operations. Analysis processes rely on information collected from hazard surveys, inspections, hazard reports, and accident investigations. This analysis process can provide a "snapshot" of hazard information. Effective analysis can then take the "snapshots" and create viable pictures of hazards and accident causal factors.

Change Analysis

Change analysis helps hazard control personnel identify hazards inherent in new processes and job-related tasks. Change analysis actually works as a proactive problem-solving technique. To solve a problem, hazard control personnel must look at situations using some type of logic process. Change analysis must attempt to identify all anticipated hazards and concerns generated by the change (Table 1.13).

Creative Hazard Analysis

Creative hazard analysis combines innovation with human expertise to identify, discover, and analyze hazards of a process, operation, or system. Ensure the hazard analysis team understands the problem statement. Provide the team with sufficient information such as known hazards, related technologies, operational procedures, equipment design issues, instrumentation controls, and necessary historical information. As the team works through each step of the hazard process, it should collectively generate a list of "what or why" questions related to hazards. After completing this list of "probing" questions, the team must systematically answer each question. This process can provide answers that can help achieve consensus. The answers can also generate additional questions that seek to clarify important information. The use of intuitive questions and answers provides insight for all team members. The team then works to achieve a consensus on each question and answer. The answers that achieve consensus form the foundation for developing recommendations or dictating the requirement for additional action or study. The team then can make recommendations to reduce operational hazards.

Table 1.13   Change Analysis Steps

- Define the problem. (What happened?)
- Establish the norm. (What should have happened?)
- Identify, locate, and describe the change. (What, where, when, to what extent?)
- Specify what was and what was not affected.
- Identify the distinctive features of the change.
- List the possible causes and select the most likely causes.

Risk Analysis

Risk analysis helps hazard control personnel assess the probability that an uncontrolled hazard could contribute to an accident event with resulting organizational losses. Risk assessments must also consider the potential severity associated with an adverse event occurrence. Analysis personnel should use available empirical data when attempting to determine probability of a risk-related event. Severity consideration should become the controlling issue when other factors indicate a low probability of an event. Risk personnel can consider hazards with acceptable risks as safe and those with unacceptable risks as unsafe. The phrase "safety first" makes a great sounding slogan and many organizations use it. Taken literally, the slogan implies that safety becomes the "primary objective" and not job or task accomplishment. However, in many very hazardous jobs and operations, a more appropriate slogan should read, "accomplish the job with safety."

Phase Hazard Analysis

Phase hazard analysis processes work very well for construction projects and other settings with rapidly changing work environments. Consider "phase hazards" as a new or unique set of hazards not present during operations. Before transitioning to a new phase, conduct an analysis to identify and evaluate new or potential hazards. Use the information gained through analyses to develop action plans that can ensure implementation of appropriate controls.

Process Hazard Analysis

The OSHA Process Safety Management Standard requires completion of a process hazard analysis for any activity involving the use of highly hazardous chemicals. The OSHA standard applies to entities using, storing, manufacturing, handling, or on-site moving of highly hazardous chemicals. Process hazard analysis permits employers to accomplish detailed studies to identify every potential hazard. The analysis must include all tools and equipment, each chemical substance, known hazards, and every job-related task. The analysis must show that each element of the process poses no hazard, contains an uncontrolled hazard, or a controllable hazard in all foreseeable circumstances. Recommend using process hazard analysis during the design and development phases of any hazardous project or operation under development.

Job Hazard Analysis

Job hazard analysis (JHA) permits the examination of job-related tasks, operations, and process to discover and correct inherent risks and hazards. Supervisors and other experienced personnel can perform the process by working sequentially through the steps of a job process or task. JHA can help tools, machines, and materials used to perform a job. JHA does require an understanding of potential job hazards. Personnel conducting the analysis must possess knowledge of hazard control including use of PPE. A well-developed JHA can also serve as an effective teaching tool. Organizations should develop a JHA for all tasks, processes, or phase-related jobs. Ensure conduct and update a JHA when a job process changes or an accident occurs. Recommend that each organization develops standardized procedures and formats for conducting the analysis. An effective analysis provides the basis for developing and implementing appropriate control measures. Post analysis results at appropriate workstations and other job or process locations (Table 1.14).

Table 1.14   Job Hazard Analyses

Step 1: Break the job down—Examine each step in the process for hazards or unsafe conditions that could develop during job accomplishments.

Step 2: Identify hazards—Document process hazards, environmental concerns, and anticipated human issues.

Step 3: Evaluate hazards—Assess identified hazards and behaviors to determine their potential roles in an accident event.

Step 4: Develop and design hazard controls—Develop or design the best hazard control based on evaluating each hazard. Coordinate implementation of all feasible controls.

Step 5: Implement required controls—Coordinate and obtain management approval for implementation.

Step 6: Revise and publish the job hazard analysis information—Update the JHA and then communicate implementation actions with the organizational members.

Job Design

Creating well-designed jobs, tasks, and processes can help reduce worker fatigue and repetitive motion stress, isolate hazardous tasks, and control human factor hazards. The concept of job design refers primarily to administrative changes that help improve working conditions. Designing safe work areas must address workstation layout, tools and equipment, and the body position needed to accomplish the job. Safe work area design reduces static positions and minimizes repetitive motions and awkward body positions. Consider the importance of human factor issues when designing work processes.

Hazard Control and Correction

Organizations must use the concept known as *hierarchy of controls*, used to reduce, eliminate, and control hazards or hazardous processes. Hazard controls can also include actions such as using *enclosure*, *substitution*, and *attenuation* to reduce human exposure risks. An enclosure keeps a hazard "physically" away from humans. For example, completely enclosing high-voltage electrical equipment prevents access by unauthorized persons. Substitution can involve replacing a highly dangerous substance with a less hazardous one. Attenuation refers to taking actions to weaken or lessen a potential hazard. Attenuation could involve weakening radioactive beams or attenuating noise to safer levels. The use of system safety methods, traditional hazard control techniques, and human factors must begin at the initial stages of any design process (Table 1.15).

Passive hazard controls would not require continuous or even occasional actions from system users. Active controls would require operators and users to accomplish a task at some point during the operation to reduce risks and control hazards (Table 1.16).

Engineering Controls

Seek to eliminate hazards by using appropriate engineering controls. Make the modifications as necessary to eliminate hazards and unsafe conditions. The design of machine guards, automobile brakes, traffic signals, pressure relief valves, and ventilation demonstrates engineering controls at

Table 1.15   Hazard Correction Monitoring System

- Implement a system to report and track hazards correction actions.
- Establish a timetable for implementing hazard controls.
- Prioritize hazards identified by inspections, reports, and accident investigations.
- Require employees to report hazards using established processes.
- Provide quick feedback about the status of hazard correction.
- Delegate responsibility for correcting and documenting completion actions.
- Permit supervisors and experienced employees to initiate hazard correction actions.

Table 1.16   Common "Never-Ever" Hazards

- Obstacles preventing the safe movement of people, vehicles, or machines.
- Blocked or inadequate egress routes and emergency exits.
- Unsafe working and walking surfaces.
- Using worn or damaged tools and equipment or misusing tools.
- Failing to identify hazards and providing proper equipment including PPE.
- Operating equipment with guards removed or bypassed.
- Permitting the presence of worn, damaged, or unguarded electrical wiring, fixtures, or cords.
- Lack of or inadequate warning, danger, or caution signs in hazardous areas.

work. For example, proper ventilation can remove or dilute air contaminants in work areas. Air cleaning devices can also remove contaminants such as particulates, gases, and vapors from the air. Using engineering, design, and technical innovation remains the top priority for controlling or eliminating hazards. Establishing preventative and periodic maintenance processes can help ensure tools and equipment operates properly and safely. Train equipment operators how to recognize and report equipment maintenance requirements and unsafe conditions. Preventative maintenance must also address engineered hazard controls and emergency equipment. If needed, schedule shutdowns to address preventative and predictive maintenance issues. Ensure the preventive maintenance addresses safety and hazard control issues as well as operational or production requirements.

Administrative Controls

Use administrative controls, such as scheduling to limit worker exposure to many workplace hazards, such as working in hot areas. However, OSHA prohibits employee scheduling to meet the requirement of air contaminant exposure limits. The scheduling of maintenance and other high-exposure operations during evenings or weekends can reduce exposures. Use job rotation to limit repetitive motion tasks or reduce the exposure time to occupational noise hazards. Use a work–rest schedule for very hazardous or strenuous tasks.

Work Practices Controls

These controls can reduce hazard exposure through development of standard operating procedures. Another important work practice relates to conducting training and education about the safe use of tool and equipment. Practices can also include knowing emergency response procedures for spills, fire prevention principles, and dealing with employee injuries. Job-related education and training helps individuals work safely and minimize hazard exposure risks. Work practice controls must address task accomplishment and ensure workers understand all job-related hazards (Table 1.17).

Warnings, Signs, and Labels

Develop and implement effective warnings and other precautionary instructions to reduce potential hazardous exposures. When posting hazard or danger warnings, consider using state-of-the-art communication methods. Affix warning signs temporarily or permanently at locations with known hazards. Tags serve as temporary signs. Tags are usually attached to equipment or a structure to warn of existing or immediate hazards. Signs and symbols must remain visible at all times where hazards exist. Use danger where an immediate hazard exists. Danger signs must use red as the predominating color for the upper panel, a black outline on the borders, and a white lower panel for additional sign wording. Use "Caution" signs only to warn against potential hazards or to take caution against unsafe practices. Caution signs use yellow as the predominating color with a black upper panel and borders. Caution signs contain yellow lettering of "caution" on the black panel and use the lower yellow panel with black lettering for additional wording.

Exit signs must contain legible red letters not less than 6-in. high on a white field. The principal stroke of the letters must measure at least 3/4 in. in width. Safety instruction signs use white with a green upper panel with white letters to convey the principal message. Any additional wording must use black letters on the white background. Directional signs use white with a black panel and a white directional symbol. Any additional wording on the sign must use black letters on the white background. Use accident prevention tags as a temporary means of warning employees of an existing hazard, such as defective tools or equipment. Do not use tags as substitutes for prevention signs (Table 1.18).

Table 1.17 Hierarchies for Controlling Hazards

- Engineering and technological innovation remains the preferred type of hazard controls.
- Substitution results in using a less hazardous substance or piece of equipment.
- Isolation moves either workers or hazardous operations to reduce risks.
- Work practices such as policies or rules can reduce human exposure to the hazard.
- Administrative controls limit human exposures through the rotation and scheduling.
- Consider personnel protective equipment when other controls prove inadequate.

Table 1.18 ANSI Safety Standards for Color Codes and Signs

- Z535.1 Color Codes for Safety Signs
- Z535.2 Environmental and Facility Safety Signs
- Z535.3 Safety Symbols
- Z535.4 Product Safety Signs and Labels
- Z535.5 Temporary Hazard Signs

Personal Protective Equipment

Consider the use of appropriate PPE and clothing when engineering, administrative, and work practice controls fail to provide adequate or mandated protection for individuals exposed to hazards and unsafe conditions. OSHA can require PPE to protect the eyes, face, head, and extremities. Examples can include protective clothing, respiratory devices, protective shields, and barriers. When employees provide their own PPE, the employer must ensure its adequacy, including proper maintenance, and sanitation. Employers must assess the workplace to determine hazards that would require use of PPE. Employers must select and require use of PPE that will protect from the hazards identified in the PPE Hazard Assessment. OSHA requires the employer to verify completion of the assessment through a written certification that identifies the workplace, certifying person, and assessment date. Never permit use of defective or damaged PPE. Train employees on the proper selection and use of PPE. Employee must demonstrate the ability to use PPE properly before using it on the job. Provide retraining whenever required for employees failing to demonstrate an understanding of proper PPE use. Never use PPE as a substitute for engineering, work practice, or administrative controls. Consider PPE as all clothing and other work accessories designed to create a hazard protection barrier. PPE should comply with applicable American National Standards Institute (ANSI) standards. Using PPE can create hazards such as heat disorders, physical stress, impaired vision, and reduced mobility. Review PPE policies at least annually. The review should include evaluation of accident and injury data, current hazard exposures, training effectiveness, and documentation procedures. The employer must verify that affected employees receive and understand required training through a written certification that contains the name of each employee, dates of training, and topics covered. Employers in most work situations must provide PPE mandated by OSHA at no cost to employees. OSHA does not require employers to pay for nonspecialty safety-toe footwear, including steel-toe shoes or boots and nonspecialty prescription safety eyewear if employees wear them away from the job-site. The employer must pay for replacement of PPE, except when the employee loses or intentionally damages PPE. When employees provide their own PPE, employers may permit use. OSHA does not require reimbursement to the employee for that equipment. Employers cannot require employees to provide their own or pay for PPE.

Eye and Face Protection

Refer to 29 CFR 1910.133 for OSHA standards covering eye and face protection requirements. Employers must provide suitable eye protection when flying particles, molten metal, liquid chemicals, acids or caustic liquids, chemical gases or vapors, potentially injurious light radiation, or any combination hazard exist in the workplace. Protective eye and face devices must comply with ANSI Z-87, Occupational and Educational Eye and Face Protection. Eye protectors must provide adequate protection against particular hazards with a reasonably comfortable fit when worn under designated conditions. Protectors must demonstrate durability and fit snugly without interfering with the movements or vision of the wearer. Finally, keep eye protectors disinfected and in good repair.

Head Protection

Refer to 29 CFR 1910.135 for information about OSHA head protection. Ensure workers wear appropriate head protection that can resist penetration and absorb the shock of blows. Evaluate the need for using protective hats to protect against electric shock. OSHA requires head protection hats to meet the requirement of ANSI Z-89.1, Industrial Head Protection, and Z-89.2, Requirements for Industrial Protective Helmets for Electrical Workers. Each type and class of head protector must provide protection against specific hazardous conditions. An understanding of these conditions will help in selecting the right hat for the particular situation.

Foot Protection

Refer to 29 CFR 1910.136 for OSHA standards addressing foot protection. Select safety shoes made of sturdy materials with impact-resistant toes. Some shoes contain metal insoles to protect against puncture wounds. Metatarsal guards can provide additional protection. Today's safety shoes come in a variety of styles and materials. Classification of safety shoes relates directly to their ability to meet requirements of compression and impact tests. Protective footwear must comply with the requirements found in the ANSI Z-41.1 Standard.

Arm and Hand Protection

Refer to 29 CFR 1910.137 for OSHA standards addressing arm and hand protection. Employers must provide appropriate protection when hazard assessments reveal engineering and work practice controls cannot eliminate injury risks. Potential hazards can include skin absorption of harmful substances, chemical or thermal burns, electrical dangers, bruises or abrasions, cuts or punctures, fractures, and amputations. Protective equipment can include gloves, finger guards, arm coverings, and elbow-length gloves. Employers must evaluate the use of engineering and work practice controls before requiring glove use. The nature of the hazard and the operations involved will affect the selection of gloves. Require employees to use gloves designed for the specific hazards and tasks.

Body and Torso Protection

Certain hazards may require the use of body protection clothing or equipment. For example, exposure to bio-hazards or chemical hazards during hazardous the mixing of dangerous drugs would require body protection. However, other hazards that could pose a risk to the body include heat sources, hot metal exposures during welding operations, hot liquids, and radiation exposures. Body protection clothing can vary and could include gowns, vests, jackets, aprons, coveralls, and full body suits. Refer to manufacturer or supplier selection guides for information on the effectiveness of specific materials against specific hazards. Inspect clothing to ensure proper fit and function.

Hazard Control Committees and Teams

Hazard control committees can make positive contributions to the success of the hazard control function. Committees can help review results of inspections, audits, and other evaluations. Top management that fails to properly assign committee responsibilities and delegate authority will undermine their effectiveness. Some top management personnel do acknowledge committee existence but ignore their contributions and suggestions. An effective committee must work hard to interact with all organizational members to improve their hazard control awareness. Top management ultimately must take responsibility for committee successes and failures. Top management

Table 1.19　Basic Hazard Control Committee Duties

- Participate in self-inspections and hazard surveys.
- Encourage organizational members to work safely.
- Assist with the job hazard analysis processes.
- Provide input for hazard control policies, procedures, and rules.
- Promote hazard control efforts at organizational meetings.
- Assist supervisors with initial accident investigations.
- Communicate employee hazard control concerns to management.

must also provide supporting systems, authority, and necessary resources to help ensure committee success. Committee membership must include nonsupervisory and hourly personnel with representation from key functions, departments, and divisions. Committee authority and responsibilities can vary depending on the organization. Some organizations refrain from using the word "committee" opting to use other terms such as "advisory panel." Regardless of the name used, committees and panels can help improve organizational hazard control efficiency and effectiveness (Table 1.19).

Hazard Survey Teams

Organizations should consider establishing a hazard survey team to supplement the established hazard control committee. This team functions best when composed of screened volunteers with no formal supervisory responsibilities. This team can help inspectors conduct walking hazard surveys to identify unsafe conditions and risky behaviors. Provide the team with a pretour education session to inform them on tour expectations and procedures. For example, if the survey will target basic electrical hazards, provide them realistic information about how to recognize and document these hazards. The team should meet after the survey to discuss and validate the findings. The organization then can implement needed corrective actions. Some team members will remember what they learned during the survey. This new information can "trickle down" to others within the organization.

Evaluation of Hazard Control Function Effectiveness

OSHA uses injury and illness rates to assess effectiveness of occupational safety and health programs. Insurance companies use an experience model to determine good and poor risks for underwriting workers' compensation coverage. Accident and injury experience does provide a good indicator about the effectiveness of hazard control initiatives. However, accident frequency and severity rates alone do not always accurately evaluate effectiveness of an accident prevention function. For example, an organization may experience an underreporting of occupational disease cases and hazardous material exposures (Table 1.20).

Placing too much emphasis on injury-producing events but not focusing on potentially serious close call incidents can result in unreliable effective assessments. Rather than relying solely on injury rates or other postevent assessments, organizations could use a broader hazard control audit process. This management style audit would address several key components of the accident prevention process. The audit forms would help evaluators rate each component against prepublished standards. Evaluators should review policies, plans, and organizational records. They should also conduct interviews, personally observe, and use employee questionnaires to help validate audit scores assigned to each component. Organizational leaders must then evaluate all other related initiatives designed to support the hazard control function. Evaluations must assess reporting accuracy and effectiveness of hazard analysis activities. Review hazard surveys and self-inspection reports to determine accuracy and comprehensiveness (Table 1.21).

Table 1.20   Symptoms of Ineffective Hazard Control

- Managers fail to recognize or acknowledge existence of hidden cultures.
- Organizational members do not view hazard control as operational priority.
- Accident investigations fail to focus on documenting causal factors and root causes.
- Senior managers fail to provide visible hazard control leadership.
- Leaders do not provide sufficient resources to support hazard control efforts.
- Leaders view hazard control as a necessary "program" but not vital to organizational success.
- Orientation, education, and training objectives fail to address "real-world" issues.
- Hazard control managers expected to prevent all accidents and injuries.
- Leaders fail to hold accountable organizational members not supporting hazard control efforts.

Table 1.21   Improving Hazard Control Effectiveness

- Use hazard analysis to define and pinpoint organizational problems or trends
- Improve the hazard control function proportionally with improvements made in other areas
- Emphasize the importance that safe behaviors and good communication play in reducing losses
- Consider hazard control as a production tool that can improve the bottom line
- Provide off-the-job safety and health education to all employees on regular basis
- Recognize the existence of covert cultures operating within the organization

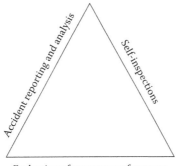

Evaluation of program performance

Evaluating the effectiveness of hazard control management programs.

Areas to Evaluate

Evaluate senior leader commitment by determining time and resources allocated to hazard control efforts. Determine which policies and procedures enhanced hazard control effectiveness. Conduct evaluations to determine how well interfacing staff functions support hazard control efforts and accident prevention initiatives. Review submitted cost-benefit analyses reports to determine accuracy and documentation reliability. If conducted, review results of perception surveys completed during the evaluation period. Review and assess the information included in organizational property damage reports. Attempt to compare or "benchmark" accident and injury data with similar types of organizations or industries. Refer to compliance and insurance inspection to ensure completion of corrective actions of all recommendations and findings. Organizations can also conduct risk measurements by reviewing historical accident and injury frequency and severity data. This retrospective measurement can provide valuable information about exposure sources that contributed to the injury or illness. Historical accident and inspection reports can also shed light on causal factors and the circumstances contributing to the injury or damage. Knowledge accident causal factors when connected to specific jobs and tasks can provide insight for recognizing potential risk.

Review Exercises

1. Why should a proactive hazard control management approach be important to any organization?
2. How does IBFCSM describe the concept of hazard closing?
3. Describe the origin of the words "program" and "function."
4. Why would the term "function" be more appropriate to describe hazard control efforts than the word program?
5. List the assumptions of "traditional" hazard control management.
6. List five key hazard control responsibilities of senior organizational management.
7. List seven key responsibilities of a hazard control manager.
8. List seven key hazard control responsibilities of front supervisors.
9. Describe the concept known as employee engagement.
10. Why should effective hazard control be considered a "good" business practice?
11. What three elements should be included in an organizational hazard control policy statement?
12. What tool does an effective emergency planner use to create an operation plan?
13. List the nine key elements of an effective hazard control plan.
14. Why should a hazard control manager develop and maintain an FHI?
15. List the five occupational hazard categories and give two examples of hazard that would fit into each category.
16. List five common factors found to be present or active in hazard-free environments.
17. Define or describe the following types of analysis:
    a. Change
    b. Risk
    c. Phase
    d. Process
18. Describe the steps involved in JHA.
19. List at least three hazard correction techniques.
20. What is the difference between work practice controls and administrative controls?
21. List at least five symptoms of an effective hazard control function.

Chapter 2

Leadership and Management

Introduction

Chapter 1 addressed the importance of effective hazard control as an integral function of an organization. Understanding some of the basic concepts related to effective management and leadership can prove valuable to those working or supporting organizational accident prevention efforts. This chapter provides a brief review of management theory and functions. It also provides a quick overview of key leadership principles that should complement management efforts. Some people think of "leadership" as just another function of management. However, I prefer to view "management" as the key support function of sound leadership. Organizational leaders and those who manage hazard control functions must consider the roles that people play in preventing or contributing to accidents and injuries. Many people view managing as an art that anyone can master with practice. Others view managing as a learned discipline or social science. Good managers seem to show that managing consists of learning and practice. The word "manager" can also refer to an individual's job or position title. How often do we meet someone with the title of manager, who shows little understanding of even basic management principles? Individuals do not become effective managers because they hold the title or position. Too often, society misuses the term manager much like they overuse of the word "safety." We tend to use both words out of context or without much thought to their true meanings.

Leading simply refers to taking actions to influence others toward attainment of organizational goals and objectives. Effective leaders understand the importance that human engagement plays in goal accomplishment. Human engagement refers to the concept where individuals personally feel their connection to their position and to organizational success. Human engagement contributes to a personal satisfaction that contributes to increased productivity, morale, and motivation.

Many organizations now realize the need for balancing organizational demands with a person's family and other life issues. Individuals, when away from the organization, serve in a variety of roles, including volunteers, caregivers, and parents. Understanding the concept of human engagement helps leaders and hazard control managers better understand the behaviors and reactions of individuals to organizational issues and decisions. Conflicting responsibilities can lead to role misunderstandings and overloads that can impact their support for organizational objectives including hazard control efforts.

Leadership

Effective leading requires the manager to motivate subordinates, communicate effectively, and effectively use power. To become effective at leading, managers must first understand their subordinates' personalities, values, attitudes, and emotions. Leaders need to identify opportunities to reward personal and team success. Place the emphasis on improving organizational systems and processes instead of blaming individuals. True leaders focus on processes and must learn to educate followers instead of dictating to them. Leaders who use "conditional statements" to encourage others must also listen closely before taking actions. Leaders must promote ownership or "buy in" of hazard control as an organizational value. Encourage creativity to increase responsible actions of those being led. However, leaders must establish and communicate expectations in clear and concise terms. Taking these actions will reduce the need for any future mandates.

True leaders learn to trust in their people skills while remaining uncertain about "how" to best meet objectives. Effective leaders learn to look beyond numbers if possible and resist trying to quantify everything. Leaders must learn to make both tactical and strategic decisions. Leaders must possess a vision of the organization structure and the path that it is traveling. System thinking helps leaders to see the "big picture" or the true organizational cultures that impact failure and success. Formal leaders must become effective ambassadors of the hazard control message (Table 2.1).

Practical Leadership

Floyd Fenix presents addresses the topic of "leading safety" very well in his commonsense booklet, *Safety Leadership Quick, Easy, & Cheap*. Fenix relies on his understanding of people and his vast experience to provide suggestions on how "true leadership" can reduce accidents and injuries. His booklet defines efficiency as the ratio between input and output. In other words, what outcome resulted when compared to the resources used? Effectiveness refers to taking the right actions to achieve a desired or expected outcome. Emotional issues can impact and even "side track" the best hazard control efforts. True leaders must learn to use logic when seeking to reduce accidents and injuries. Fenix indicates that people responding to an event or situation make initial decisions based on emotions. Less than 10% of the time individuals make initial decisions using a logic approach. The appropriateness and quality of decisions at every organizational level can impact hazard control effectiveness. Fenix indicates that most individuals will retain only about 11% of what they hear one time. However, retention can improve to more than 90% after hearing the

Table 2.1 Concepts Leaders Must Understand

- Character refers to moral or ethical structure of an individual or group.
- Belief refers to the mental act or habit of placing trust in someone or something.
- Value refers to an individual's perception of worth or importance assigned to something.
- Culture reflects the socially accepted behaviors, beliefs, and traditions of a group.
- Attitude refers to an individual's personal state of mind or feeling about something.
- People can perform a job or task and hide their attitude from others.
- Behavior relates to an open manifestation of a person's actions in a given situation.

Table 2.2   Human Senses and Learning New Behaviors

Human Sense	Effectiveness for Learning (%)
Sight	87
Hearing	7
Touch	3
Taste	2
Smell	1

Source: Table created from information obtained from Fenix, F., *Safety Leadership, Quick, Easy and Cheap*, Malvern, AR: EZ Up Inc, 1999.

same information for 11 times. Leaders should understand that repetition acts as the "mother" of learning. Leaders should also consider human learning abilities, including retention of information, when developing orientation, education, and training session. Chapter 3 contains additional information about education and training methodologies.

Fenix stresses in his text that the numbers of "poor leaders" considerably exceed the number poor employees. Improved leadership results in better employees. According to Fenix, leaders must understand that recognition, appreciation, and praise serve as the top three motivators of people, and good leaders use positive reinforcement about 95% of the time. Hazard control promotes the importance of recognizing and identifying unsafe work conditions and behaviors. Fenix stresses the importance of improving "worker awareness" of hazards and exposures encountered on the job. Supervisors should continuously stress the need for improving awareness on the job. High reliable organizations that use system safety methods place a strong emphasis on both task and situational awareness on the job. According to Fenix, getting individuals to replace bad or unsafe habits with proper or safe habits will improve safety-related behaviors in the workplace. Someone must remind the average person at least 11 times not to engage in a bad or unsafe habit. Some, then, must remind the same person another 11 times to develop the good or safe habit (Table 2.2).

Ethics and Leadership

When addressing leadership, consider ethics as vital since it provides the foundation of any hazard control management function. Without ethics, hazard control loses its organizational and personal value. Webster's Collegiate Dictionary defines ethics as (1) "discipline of dealing with what is good and bad and with moral duty and obligation," and (2) "a set of moral principles or value or a theory or system of moral values." Ethics finds its roots in natural law, religious tenets, parental or family influence, educational experiences, life experiences, cultural norms, and societal expectations. Business ethics refers to the application of discipline, principles, and ethical theories to the organizational context. We can define business ethics as the "principles and standards that guide behaviors in the business world." Ethical behaviors remain an integral part of conducting business affairs. The three considerations that impact and influence ethical decision making in business include individual difference factors, operational or situational factors, and issue-related factors.

Management

It seems so easy create and use the phrase "safety management" as if communicated a definitive concept or process. However, safety management can mean different things to different people. The National Safety Management Society defines the term as "… that function which exists to assist all managers in better performing their responsibilities for operational system design and implementation through either the prediction of management systems deficiencies before errors occur or the identification and correction of management system deficiencies by professional analysis of accidental incidents (performance errors)." This definition stresses the importance of identifying and correcting management-related deficiencies. Many people never think of addressing management deficiencies when referring to the term "safety management." Poor and inefficient management provides opportunities for accidents to occur. Some individuals also think of management and leadership as synonymous terms. You can manage materials, projects, and processes, but you must lead people.

Organizations need all managers and supervisors to provide hazard control leadership to their subordinates. I often made this statement while teaching safety educational sessions: "Good leaders who can't or don't want to be burdened with the managerial details must quickly find someone who can help them." I then follow up with this statement, "Good managers who can't lead or don't want to lead need to quickly learn the art of delegating to someone who can." Dealing with too many details can cause operational managers to become overburdened with objectives, goals, and time constraints.

Many organizations now use "project management" personnel to oversee the activities of long-term or complex projects. Leaders must move toward meeting established objectives, but do so by considering both people and process during the journey. Hazard control management efforts without leadership can easily fail. However, hazard control professionals not using good management techniques can also fail in meeting expectations or objectives. Managers must set an example by following work rules and behavioral expectations established for their subordinates. Organizational members must see management personnel consistently setting a positive example in the area of hazard control. Managers must make practicing safety a priority and lead by example. Managers must learn to help other achieve personnel success while meeting established organizational objectives. People perform better when provided with proper information, necessary tools, and delegated authority to get a job done. People must view themselves as participants in a project and never a pawn of a manager.

Management Theories

Properly using the functions of management can contribute to the success of every organized enterprise. Traditional management functions include activities such as planning, organizing, coordinating, directing, and controlling. The functions of the management work together in a synergistic process to improve organizational operations including hazard control effectiveness. The evolution of organizational management theories began to take hold in the early 1900s with Frederick Taylor's scientific management theory. Taylor advocated finding the single best way to perform a task and then select the best person to accomplish it.

During the 1930s, Mooney and Reiley proposed a set of standard management principles known as administrative theory. They proposed that a standard approach could apply to all organizations. During the 1940s, Max Weber began to promote his bureaucratic theory based on a hierarchical power structure. Chester Barnard defined organization as being a system of consciously coordinated activities. Barnard also stressed that senior leaders should create an atmosphere of value and purpose. Alfred Chandler studied large corporations and suggested they would go through some type of evolution to meet their strategy, mission, or function. Lawrence and Lorsch suggested that top leaders delegate authority to lower managers to improve decision-making effectiveness. The need for a better understanding of organizational management led to many proposals, studies, and theories beginning the 1950s. The emergence of system thinking, which considers all components as interrelated, helped leaders better understand complex processes. The use of system approaches also helped managers better comprehend how a single variable within a process can impact or cause changes in other variables within the same process. Henri Fayol and Frederick Taylor suggested that contingency or situational management could deal with inevitable conflict that would arise within organizations as they became more complex. Fred Fiedler suggested, in the 1960s and 1970s, the key determinants of a leadership situation depended on the degree to which a subordinate trusted a leader, the formal authority held by a leader, and the degree of structure of a job task.

Peter Drucker

Drucker proposed during the 1970s, the concept of decentralization for large organizations. Drucker suggested creating different organizational divisions that would operate simultaneously. He also suggested that some division would operate independently. This would permit organizations to diverse by allowing dynamic but flexible operations. However, he also believed that this divisional independence would hinder the integration and coordination of organizational activities. Today, many large corporations and companies must address the reality of Drucker's foresight about decentralization. You cannot operate a traditional or centralized hazard control function in an organization that operates in a decentralized manner.

Classic Management Theory

The classic theory of management proposed by McGregor can play an important role in hazard control and can help prevent accidents within some organizations depending on organizational structure. Theory "X" or the traditional management approach develops mandatory rules and policies enforced by the organization. Theory X can work in centralized, bureaucratic, and "line" organizations. Theory "Y," on the other hand, promotes worker involvement and participation in the functions of the organization. A good example of a Theory "Y" approach would be to include nonsupervisory hourly employees on the facility hazard control committee. Managers should look for ways to motivate, educate, and encourage workers to work safely. Theory Y approaches would work better in decentralized organizational environments.

Motivation-Hygiene Theory

In the late 1950s, Frederick Herzberg's promoted his motivation-hygiene theory of leadership. The theory proposed that people require different things to motivate and satisfy their needs. Herzberg proposed that some factors contributed to job satisfaction but other factors did not. He also noted that some factors served as a source of dissatisfaction when not present. However, dissatisfying factors did not function as motivators. Motivating factors go the very root and nature of an individual's job or position. Hygiene-related factors do little to motivate people but can cause great dissatisfaction when not provided or present. Job security could serve as a motivating hygiene factor, while recognition would result in motivating the individual. Organizational hazard control could serve as a hygiene factor in organizations that view accident prevention as just another program. However, hazard control could serve as motivator in organizations that view accident prevention as integral part of every job, task, or process.

Managerial Grid

Robert Blake and Jane Mouton in the 1960s created the managerial grid to plot the degree that a manager focuses on people or job accomplishment. The grid considers a manager's concern for the people who do the job and his or her concern for the job itself. The grid uses an axis to plot managerial concerns for job completion versus concerns for the needs of people. Blake and Mouton defined the following five basic managerial styles. When using the Managerial Grid, consider the process as only a starting place for seeking a better understanding of management commitment categories. To perform at their highest level, hazard control managers must develop a good understanding of basic management and leadership concepts (Table 2.3).

Theory S

William C. Pope, in his excellent book entitled, *Managing for Performance Perfection: the Changing Emphasis*, presents several principles of system safety management or Theory "S." Pope's Theory "S" supports organizational performance and quality objectives by stressing the need for direct human involvement in hazard control functions. Hazard control functions must discover and document poor performance to show that the costs associated with errors does not result in too great of expense to the margin of profit. Pope believed that management should control the behavior of any system. He proposed that shifting this responsibility, to assign personal blame, did not correct the human or interpersonal situations that permitted a mishap to occur repeatedly. Administering any system, to avoid flaws, required adapting the system to its environment. This adaptation ensured management of the safest possible performance of both human and material resources. Pope suggested that three "system" generated flaws or causal factors exist in all organizational mishaps. He referred to the flaws as biological or human, physical or property, and social or management. He believed that correcting flaws in one system and not addressing the other two systems can increase potential for additional mishaps. Management must take responsibility for the flaws in all three systems.

Table 2.3   Blake and Mouton Managerial Styles

High people/low job: This type of manager believes if you take care of people they will work hard at their jobs. This type of manager does not provide workers with direction and job accomplishment suffers.
High job/low people: This type of manager believes that the job always remains the top priority. Managers of this type flourish in organizations that use punitive actions to motivate people.
Low job/low people: This type of manager serves in title only with little regard or concern for job or the people hired to accomplish it. We can categorize this type of person as a "leech" who collects a salary while feeding on failure.
Medium job/medium people: This type of manager attempts to balance his or her concern for both the job and people by compromising. We can categorize this management style as "don't rock my boat" while we sail on the "sea of mediocrity."
High production/high people: This type of manager also performs as a leader. He or she places a high priority on producing results and the understanding the needs of the people performing the tasks. We can categorize this management style as "leadership and teamwork."

Theory Z

Proposed by Professor William Ouchi, Theory "Z" addresses long-term job security, consensual decision making, slow promotion opportunity, and individual responsibility within a group context. Theory Z breaks away from McGregor's Theory "Y" that focuses on employer–employee relationships. Theory "Z" changes the level of to include entire organizational structure. Theory "Z" characterizes values, beliefs, and objectives as similar to "clannish" cultures. The theory places an emphasis on the socialization of group members for the purpose of achieving personal and group objectives. Organizations can retain some components of a bureaucracy including authority delegation and evaluation of performance. Some view Theory "Z" as suggesting that common cultural values promote increased organizational commitment among employees.

Learning Organizations and Managing Knowledge

In the early 1990s, Peter Senge promoted the learning organization concept in his book, *The Fifth Discipline*. He defined learning as improving your ability to act. He believed it was a serious organizational issue when individuals viewed themselves as a part of their job position. Overidentifying with a position resulted in organizational members not comprehending their role in the larger organizational system. He also suggests that many management teams comprising skilled individuals will actually protect themselves from outside threats if placed on such a team. He found that the reaction to being on a management team actually caused these skilled people to distance themselves from the process of learning. Organizations can improve learning by adopting the following "Senge disciplines" that contribute to learning. The first discipline deals with sharing and committing to the organizational vision. Next, team members must build their own mental models of reality and share them with others. Finally, team members must develop a commitment to a "system approach" to organizational success.

The concept of "knowledge management" goes beyond the concept of informational technology. Many organizations manage information but neglect to manage knowledge. Managing knowledge requires the identification, analysis, and understanding of all operational processes. Knowledge provides no organization value unless the information is relevant, available, and disseminated to end users. The failure to communicate accident and hazard information can result in poor analysis of hazards or accident experience. Each organizational process, department, or function must contribute information about accidents, hazards, and unsafe behaviors. This aggregate knowledge provides the basis for determining appropriate hazard controls, the need for training or education, or required innovations to improve performance (Table 2.4).

Table 2.4 Improving Knowledge Management

- Identify the knowledge that the organization already possesses.
- Determine the kinds of knowledge that the organization needs.
- Evaluate how knowledge can add value to organizational effectiveness.
- Create processes to help the organization to achieve objectives.
- Maintain and effectively using knowledge assets to improve the organization.

Decision Making

People use management concepts to help them make good decisions. Managers must constantly evaluate alternatives and make decisions regarding a wide range of matters. Decision making involves uncertainty and risk. Many decision makers possess varying degrees of risk aversion when making decisions. Decision making may require evaluating information and data generated by qualitative and quantitative analyses. Decision making must rely not only on rational judgments but also on factors such as decision-maker personality, peer pressures, organizational situations, and a host of other issues. Management "icon" Peter Drucker identified several key decision-making practices that successful executives used. Before making decisions, answer the following questions. The first question simply says, what needs accomplishing? The second question seeks to find out, "what's best for the organization?" When a decision maker gets the answers, he or she can then proceed with developing a plan of action. Decision makers must take responsibility for their actions. True decision makers use the team pronouns such as "we" and never the self-gratifying pronoun "I."

Management by Exception

Management by exception (MBE) enables managers to quickly make common day-to-day decisions. MBE would prove inappropriate for decisions requiring significant deviation from established practices, standards, or expectations. Managers of today can easily use MBE decision-making principles because of their immediate access to electronically provided sources of information. Managers only need to be provided with sufficient information about a situation to make a decision.

The use of MBE concepts can encourage subordinates to use judgment in deciding which situations or concerns should be brought to management's attention. MBE motivates subordinates to work within established controls and reduce the number of insignificant decisions made by managers. MBE helps direct a manager's attention to issues, challenges, or problems that seriously impact organizational success. Time and effort should never be wasted focusing on those parts of the organization where things are going smoothly.

Functions of Management

Henri Fayol proposed his five management functions in his 1916 book titled *Administration Industrielle et Generale.* Fayol, who managed a French coal-mining operation, defined planning, organizing, commanding, coordinating, and controlling as key managerial functions. Fayol proposed that these functions universally applied to all managers. His view of management functions greatly impacted modern management study that began in the 1950s. The "process school of management" became a dominant paradigm for studying management and its functions. The functions became the common way of describing characteristics of management. However, Fayol's original functions did not "directly" address the "informal relationships" that could exist among managers, subordinates, and other organizational members. Many professionals now agree that "the coordinating function" must operate in all other functions of management in an integrated manner. Some management experts also like to add "staffing" to the list of functions. Fayol's "function theory" might not address many of the challenges managers face today, but his work provided a structure on which to build a management foundation.

Planning

Proper planning can help any organized enterprise in successfully meeting objectives and attaining goals. The function of planning focuses on developing a course action and setting organization direction as necessary to achieve desired outcomes. Planning requires that managers learn to make good decisions. Tactical organizational planning refers to creating and developing short-term or specific ways to implement strategic or long-term goals. Strategic planning must identify long-range opportunities and threats.

When developing tactical and strategic plans, effective managers must understand the strengths and weaknesses of their organization. Planning provides the foundation to ensure the success of all functions of management. Effective planning must consider issues related to properly use human and other material resources. Effective planning must seek to avoid confusion, uncertainty, and human emotion hindrances to goal accomplishment. Leaders must also consider organizational culture and climate during planning sessions.

Directing

Managers must effectively supervise others and attempt to motivate them to achieve success in pursuit of organizational objectives. The function and art of directing people must using effective communication and human relation skills. Directing remains the fundamental aspect of all effective management. Leadership plays an important role in directing because it can encourage and motivate others to succeed. Directing also requires overseeing the work of others, including providing them guidance and incentive.

Organizing

Managers must organize the workforce and materials in an efficient manner. Organizing includes designing the structure and aligning the activities of the organization. Organizing can also include what some refer to as the "staffing function" of management. Organizing focuses on people, activities, and other functions operating at a high level of effectiveness and efficiency. Organizing can involve designing jobs and tasks to maximize operational performance. Organizing must also consider using material and financial resource as required to ensure action plans lead to goal attainment. The art of organizing requires the use of good coordination to "connect the dots" of organizational dysfunction that would otherwise hinder goal accomplishment. Organizing efforts must identify and classify the many activities necessary to ensure success. Organizing also must address requirements related to assignment of duties, delegation of authority, creation of responsibility, and obtaining approval or consensus.

Controlling

Controlling involves the processes related to ensuring operational activities adhere to organizational policies, procedures, and directives. Managers at all levels must learn to recognize and report deviations from plans or objectives. Controlling also includes taking necessary actions to correct performance and actions that deviate from standards and expectations. Performance standards can address financial issues such as revenue, costs, and profits. Controlling can also consider other quality or production concerns such as defects or customer complaints. The controlling function must also attempt to predict substandard deviations and accomplishments before occurrence. Poor controlling functions can hinder goal accomplishment.

Coordinating

Coordinating goes beyond simple communication and feedback processes. Coordinating seeks to attain consensus among interested or relevant parties about an action or objective. Managers must learn harmonize procedures and activities performed by various functions, departments, leaders, and subordinates to ensure the organization moves forward with agreement and understanding. Coordinating, by nature, must bring functions, cultures, and groups together for the common good of the organization. Coordinating focuses on unifying goal attainment efforts. Coordinating, the hidden force, binds together other management functions. For example, directing requires coordination since there must be rapport between a superior and subordinate.

Staffing

Many business and management professionals now include "staffing" as a key function of management. Staffing deals with organizational structures and ensures the maintenance of a proper level of manning. Some organizations consider staffing as vital because of increased technology, wide diversity, and complexity of modern organizations. Staffing must consider issues such as recruitment, screening methods, assessment, proper classification, and effective assigning of people to organizational positions and departments.

Improvement Management

LEAN

Using "LEAN" processes can improve accident prevention and hazard control efforts in some organizations. LEAN safety efforts require the focus to remain in improvement and not compliance. LEAN safety efforts use a proactive approach that requires stakeholders to participate in the process by accepting responsibility for hazard control or safety improvement. The textbook, *Lean Safety: Transforming your Safety Culture with Lean Management* (CRC Press 2009) written by Robert Hafey provides an excellent read for organizations desire to build a true hazard control or safety culture. Safety culture development can help remove fear from the accident investigation process to permit discovering of root causes.

SIX SIGMA and Safety

This popular methodology attempts to improve performance by measurements that relate to quality. The process focuses on productivity, cost reduction, and quality improvements. The process uses measurement tools to enhance the use of reengineering and innovation improvements. The process does require an organizational culture change to ensure success. The success of the process depends on using effective, problem-solving tools along with process optimization methodologies. This process can help address specific accident- and injury-related operations or issues. The process requires total commitment of all involved parties. Controlling a hazard or avoiding a risk can reduce accidents and injuries. SIX SIGMA processes can help organizations "define the hazard control or safety related problem" that needs addressing. Next, determine the scope and identify "specific" measurable objectives for the injury-producing hazard or operation. SIX SIGMA can use available historical information, injury statistics, and documented root causes to help team members intuitively assess operational risks and hazards. This can lead to recommendations that would prevent accidents and incidents in the process under study.

Project Management

"Project management" involves applying leadership principles and management concepts to a transition, phase, or temporary endeavor requiring great attention to detail. This type of management can trace its roots to construction and other complex undertakings. Trained managers address and coordinate all the details related to the myriad of activities of a given project. Many project managers possess some experience in addressing quality, risk, and hazard control issues. When properly addressing these issues, projects can progress as planned and also meet budgetary or resource expectations. Most project managers use a "structured problem solving approach" to address the risk and hazard control-related issues. Hazard control personnel should coordinate with the project manager issues such as objectives, scope, time frames, costs, regulatory issues, and required performance measures.

Psychological Safety

Workplace-related psychological safety shows itself when employees feel unable to put themselves on the line, ask questions, seek feedback, report problems, or propose a new idea without fearing negative consequences for themselves, their jobs, or their careers. A psychologically safe and healthy workplace actively promotes emotional well-being among employees while taking all reasonable steps to minimize threats to employee mental health.

Crisis Management

The study of crisis management originated with the large-scale industrial and environmental disasters in the 1980s. Three elements common to most definitions of crisis includes determining threats to the organization, planning for the elements of surprise, and making decisions in short time frames. The fourth element relates to addressing the need for change. When change does not take place, some could view the event as a failure or incident. Crisis management uses response methods that address both the reality and perception of crises. Organization should develop guidance to help define what constitutes a crisis and what triggers would require immediate response.

Organizational Dynamics

Traditional Organizational Structure

Traditional organizational "structure" follows two basic patterns. The first structure, referred to as a line organization, permitted top management to maintain complete control with a clearly defined chain of command. This basic line structure works well in small companies with the owner or top manager functioning at the top of the organizational structure. Everyone understands the clear "lines" of distinction between the owner or manager and subordinates. A line-and-staff organization combines the line organization with appropriate staff departments that provided support and advice to the line functions of the organizations.

Many medium and large organizations use the line-and-staff structure, with multiple layers of staff managers supporting overall operations. An advantage of the line-and-staff organizational structure relates to the availability of technical and managerial functions. The organization incorporates these needed staff and support positions into the formal chain of command. However, conflict can arise between line and staff personnel creating disruptions within the organization. This conflict can, at times, impact the effectiveness of the hazard control management function. Hazard control managers must remain focused on identifying and correcting the causes of accidents regardless of the organizational structure. However, they must understand that organizational structure can hinder accident prevention efforts. Organizational leaders must integrate the function of hazard control into the organizational management structure with clearly defined responsibility and authority.

Top management must focus on identifying and correcting operational and staff management-related deficiencies that could hinder hazard control efforts. The organizational structure must consider developing processes that help identify and analyze system deficiencies that contribute to accidents. Leaders must ensure that support functions such human resources, facility management, and purchasing receive information about their management deficiencies that could impact hazard control.

Consider the following scenario: a human resource department mistakenly assigned a new employee to a hazardous job position without properly screening or evaluating the person's qualifications. This could contribute to an accident or mishap. Senior leaders of staff and departments must understand their roles and responsibilities related to hazard control. Many organizational structures permit and even unknowingly encourage support or staff department managers to create their own "little dynasty." This can result in the self-coronation of "turf kings and queens." Once crowned, these rulers may not see the need to coordinate or communicate important issues with other functions.

Organizational Culture

Organizational culture consists and exists based on assumptions held by a particular group. These assumptions can include a mix of values, beliefs, meanings, and expectations held in common by its members. Cultures can determine acceptable behaviors and problem-solving processes. Organizational trust refers to the positive and productive social processes existing within the workplace. Trust can encourage group members to engage in cooperative and expected organizational behaviors. Trust also provides the foundation for showing commitment and loyalty. For example, an organization with a safety-and-health-focused culture enhances the well-being, job satisfaction, and organizational commitment of all members. A culture with social support systems can enhance member well-being by providing a positive work environment for those dealing with depression or anxiety. The established culture sets the tone for an organization. Negative cultures can hinder the effectiveness of the most well-designed policies and procedures. Unhealthy cultures create stressful environments that can lower employee well-being and impacts organizational productivity (Table 2.5).

Covert and Overt Cultures

Many times, senior managers fail to acknowledge the existence of the covert, informal, and hidden cultures. They incorrectly hold the belief that the established overt, formal, or open culture drives organizational success and productivity. They also fail to acknowledge the tremendous influence

Table 2.5 Elements for Creating Safety Cultures

- Positive perception of teamwork
- Safe behaviors exist as the norm
- Job satisfaction
- Perception of senior management effectiveness
- Recognizing the reality of job-related stress
- Adequacy of supervision, education, and training
- Opportunities for effective organizational learning
- Nonpunitive response to error by leaders

of hidden cultures on organizational behaviors. Why do these hidden cultures exist? They exist to meet the needs of its members. Failing to acknowledge these hidden cultures can hinder an organization's ability to change or improve the formal culture. The actual climate, not the established structure, exerts the most influence on organizational performance. Hidden cultures can also support a very effective organizational communication process known as the grapevine.

The Grapevine

The grapevine serves as the informal and confidential communication network that quickly develops within any organization to supplement the formal channels. The actual function of the grapevine will vary depending on the organization. For example, it could communicate information inappropriate for formal channels. The grapevine can carry both good and bad organizational news. The grapevine, in some instances, serves as a medium for translating top management information into more understandable terms. The grapevine also serves as a source of communication redundancy to supplement formal channels. When formal communication channels become unreliable, the grapevine can quickly operate as the more trusted communication system.

Culture Socialization

Some organizations use socialization processes to educate new members about the organization's cultures. Socialization can occur in both formal and informal culture arenas. The socialization process will determine which culture, formal or informal, will exert the most influence on an individual. Organizational members then must decide to remain or leave the organization. Many who stay may experience isolation.

Culture Change

Leading culture change requires a sound understanding of organizational behaviors, attitudes, expectations, and perceptions. Changing the organizational culture must also impact the behaviors, attitudes, and perceptions of all organizational members. Leaders must use all available sources to communicate the "change" message. Change can have both a negative and positive impact on leaders as well. Change can dethrone the turf kings and queens. However, it can encourage development of teamwork and continuous improvement processes. When leading change within any organization, never forget the importance of communication and feedback throughout the entire journey. Most people naturally resist change that causes organizations to change very slowly. First-level change deals with people, structure, policies, and procedures. Second-level change deals with complex systems, cultures, and processes. Many people now view change as an inherent and integral part of organizational life. Some new trends in organizational dynamics emerged during recent years. These emerging trends can create conflict and concern for organizational leaders and members. The trends can create both opportunities and threats in the minds of people.

Any change creates tensions that leaders must address to prevent unwanted or dysfunctional change results. Many organizations now operate on a global scale with increased competition. These organizations also must embrace economic interdependence and increased collaboration. This globalization results in a wide range of consumer needs and preferences.

Change does not occur just because a top manager writes a memo and declares it done. Organizational change must consider how to best transform the organization from within. When leaders communicate need and intent to change, they must provide information to addresses issues such as behaviors and expectations. Communicate the reasons for change so that everyone understands. Provide guidance and leadership action to ensure change happens. Before change can occur in large organizations, leaders must acknowledge the existence and influence of hidden or covert cultures operating in the organization. Always articulate how change will impact all organizational members and operational functions. When educating others about change, provide sessions that focus on real situations. Work to change or shift the culture before implementing other innovations or interventions. Not to do so would provide no supporting foundation for the innovations or interventions. Leaders must promote trust and ensure the involvement of organizational members in decision-making processes. Providing team members with the opportunity to "voice" their concerns or suggestions can help build trust. Encourage team members to express some kind of choice and allow them flexibility to make decisions related to their job tasks. Migrating decision making permits individuals with the appropriate expertise, education, or experience, regardless of rank or position, to make informed decisions.

When changes in work sites, processes, materials, and equipment occur, hazard can emerge. During any change process, hazard surveillance and self-inspections must become frequent. Organizations should develop an enhanced hazard review process when undertaking major changes in processes, systems, or operations. Maintain sound coordination and communication systems among all parties involved in the change process.

Change may require revising existing job hazard analyses, reviewing standard operating practices, evaluating lockout methods, and assessing personal protective equipment requirements. Change-related hazard analyses can prove cost-effective in terms of preventing accidents, injuries, and other organizational losses. Individuals respond differently to change. Some organizational members may require additional time adapt and accept the change.

Effective Speaking and Writing

Hazard control managers must learn to understand the communicative process and show their ability to speak and write effectively. Communication consists of the sender, the message, and the audience. For communication to be successful, the audience must not only get the message but also must interpret the message in the sender's intended way. Communication refers to the purposeful act or instance of transmitting information using verbal expressions or written messages. To function effectively, all leaders and managers need to know how to effectively communicate with all organizational members. Managers and leaders must understand the different communication channels available.

Downward communication involves more than passing information to subordinates. It can involve managing the tone of the message and effectively showing skill in delegation. When communicating upward, tone becomes more crucial along with timing, strategy, and audience adaptation. A sender wants to transmit an idea to a receiver through using signs capable of perception by another person.

Communication

Communication refers to sharing or exchanging of thought by oral, written, or nonverbal means. Communication signs can include the printed or spoken word, a gesture, a handshake, or a facial expression. The receiver takes those signs, interprets them, and then reacts with feedback or simply ignores the message. When communicating, a sender encodes a message using some tangible sign. A sign may consist of anything seen, heard, felt, tasted, or smelled. The receiver decodes the message to comprehend its meaning. The meaning of message can differ since both the sender and the receiver can assign their own meanings. Each individual's unique set of experiences can function as a perceptual "filter." The filter blends the education, upbringing, and life experiences of the perceiver. Leaders must learn to use effective communication to provide vision and direction to others. Motivating, inspiring, and persuading others to work together require effective communication skills. Miscommunication can result in expensive mistakes, organizational embarrassment, and in some cases, accidents or even death.

Today, communication effectiveness can suffer from too much information. Around-the-clock media coverage, e-mails, and web-based informational sources make it difficult to filter the valuable information need to accomplish our goals and objectives. We must learn to communicate with clarity and focus. Failure to communicate relates to answering the wrong question, answering only part of the question, and adding irrelevant information. Many communicators answer the question but provide unnecessary feedback information.

Communication Barriers

Communication barriers, often also called noise or static, can complicate the communication process. Although unavoidable, both the sender and the receiver must work to minimize them. Interpersonal communication barriers can arise within the realm of either the sender or the receiver. If an individual holds a bias against the topic under discussion, anything said in the conversation can affect perception. Organizational barriers can occur because of interactions taking place within another larger work unit. The serial transmission effect takes place when a message travels along the chain of command path. As it moves from one level to the next, it changes to reflect the person who passed it on. By the time a message travels from bottom to the top of the chain, it changes and would not likely be recognized by the person who initiated it. Nonverbal communication occurs when information exchanges through nonlinguistic signs. Many consider "body language" as synonymous with nonverbal communication. Body language provides a rich source of information during interpersonal communication. The gestures that people make during an interview can emphasize or contradict what they say. Posture and eye contact can indicate respect and careful attention (Table 2.6).

Effective Writing

When preparing to communicate using written correspondence, organize information using a logical and systematic process. This helps the recipients to understand the message without reading it over and over. When communicating in clear manner, place emphasis on the rules of language,

Table 2.6 Focus Principles of Effective Writing and Speaking

- Focused: Address the issue, the whole issue and nothing but the issue.
- Organized: Systematically present your information and ideas.
- Clear: Communicate with clarity and make each word count.
- Understanding: Know your audience and its expectations.
- Supported: Use logic and support to make your point.

Table 2.7 Seven Steps for Effective Communication

- Analyze purpose and audience.
- Research your topic.
- Support your ideas.
- Organize and outline.
- Draft.
- Edit.
- Get feedback and approval.

including proper spelling and correctly pronouncing words. Good communicators also learn how to assemble and punctuate sentences. Effective communicators never hide their ideas or information in a jungle of unnecessary verbiage. Use of incorrect language can cripple credibility and limit acceptance of ideas. Developing strong language skills require commitment. Many writers and speakers cripple their attempt at communication by using bureaucratic jargon, big words, and too much passive voice. Good writers and speakers want to inform or persuade their audience. Building credibility with the targeted audience requires the use of support and logic. Nothing can cripple a clearly written and correctly punctuated correspondence quicker than a fractured fact or a distorted argument. However, properly using logic remains a challenge for many individuals to master, since it challenges the mind's ability to think in the abstract. Never attempt to hide intellectual shortcomings with verbal overdose. Communicators need to consider a seven-step approach that will support communication success. Good communication requires preparation, and the first four steps lay the groundwork for the drafting process (Table 2.7).

Purpose for Writing

Too many writers launch into a project without a clear understanding of their purpose or audience. Carefully analyzing your purpose helps answer the question, the whole question and nothing but the question. Take time to understand your audience, and consider their current knowledge, interest, and motives. Abraham Lincoln said it well: "Truth is generally the best vindication against slander." Support your communication with information relevant to your point. Do your homework to get educated about your communication topic.

Persuasive Writing

Since communication involves persuasion, use information that supports a logical argument. People use logic to make decisions and solve problems. Before starting to write or speak, organize your thoughts and develop a presentation outline. Good communicators organize their information logically and lead their audience from point to point. Audiences tune out speakers or writers who ramble without any logical pattern. Organize the presentation to help the audience better understand the point. The first four steps of the FOCUS process apply to both writing and speaking. Mastering the first four steps will make the actual drafting process less painful and more efficient. There are several practical points that help connect with your readers. Make your point quickly, organize the paragraphs to lead the reader, and use proper transitions to guide them. Write clear and direct sentences that cut through the jargon.

Editing

Good writers resist the laziness of overusing the passive voice. Select the right words and summarize your message in a concluding paragraph. Remember the inevitability of criticism and judgment. Learn to critically evaluate and correct your own writing. Learn the two most critical aspects of the editing process. Know what to edit and how to edit. The "how to edit" poses the biggest challenge to most writers. Start with the message as a whole and then work down to details such as spelling and punctuation. Most individuals show a limited inability to criticize their own work. Sometimes engaging someone else can help writers see how to improve or strengthen their communication. Never let pride or fear of criticism hinder objective and helpful criticism.

Writing a Report

An effective report should include a brief summary, introduction, findings, conclusion, and recommendations. The summary should include a very brief digest of the report sections. The introduction should include the need for the report and who wrote it. The main part should contain the findings, with a discussion and analysis of each finding. The conclusions and recommendations should contain statements for corrective actions including alternatives or options. Writing a report to encourage hazard control decision making or management actions can help senior leaders address important organizational issues.

Review Exercises

1. Explain in your own words how the "grapevine" can impact an organization and its members.
2. What reasons can you give that would support the author's view that effective management is a function of leadership?
3. What reasons can you give that would support the assertion that "leaders need to educate more and dictate less?"
4. Explain in your own words the difference between effectiveness and efficiency.
5. Why is ethics so important to the practice of hazard control?
6. In your opinion, what aspect of "Theory S" would be most relevant to organizational accident prevention efforts?
7. In your opinion, what are the three most important "leading safety" concepts made by Floyd Fenix?
8. Why should leaders understand the difference between effective knowledge management and information management?

9. List and define five traditional functions of management.
10. Define organizational culture and then explain the key difference between overt and covert cultures.
11. Why would effective oral and written communication skills be important to the success of a hazard control manager? How could the lack of communication skills impact hazard control efforts?
12. What are the four elements vital to effective crisis management?
13. What type of approach do most project managers use to address hazard control issues?
14. Explain Peter Drucker's concept of organizational decentralization.

Chapter 3

Understanding Accidents

Introduction

We can simply define an accident as "an unplanned event that interferes with job or task completion." When an accident occurs, someone will lose valuable time to dealing with the event. An accident can result in some kind of measureable loss such as personal injury or property damage. An accident event can also be classified as a "near miss" with no measurable loss. Accident causes normally result from unsafe acts, hazardous conditions, or both. Accident prevention efforts must emphasize the importance of developing necessary policies, procedures, and rules. The hazard control plan should outline organizational objectives, goals, and responsibilities. The organization needs to evaluate the priority and effectiveness of accident-prevention efforts.

The costs of accidents should provide motivation for senior leaders to support hazard control efforts. Accidents resulting in injuries or property damage can cause interruption of production or other operations. Hazard control managers must endeavor to obtain management's attention and support by communicating to them losses in terms of dollars and manpower utilization. We can calculate or closely determine the direct costs associated with an accident. However, determining indirect costs can pose a challenge to the best managers and hazard control managers. Traditionally, most hazard control and safety personnel held the view that indirect costs of an accident far exceed the calculated or known direct costs. Fred Manuele's thought-provoking article entitled "Accident Costs: Rethinking Ratios of Indirect to Direct Costs" appeared in *Professional Safety* in January of 2001. His article encouraged safety management to refrain from using any ratios that data could not accurately support. He wisely pointed out that the direct costs of accidents did increase significantly in recent years because of indemnity and medical expenses (Table 3.1).

Accident Causation Theories

Henri Heinrich's Five-Factor Accident Sequence

Heinrich's research in the area of accident causation concluded that 88% of "investigated" accidents resulted from unsafe acts. Heinrich attributed 10% to unsafe conditions, and he classified 2% as unpreventable. Heinrich suggested that an individual's life experiences and background could "predispose"

Table 3.1   Common Myths about Accidents

- Accidents result from a single or primary cause.
- Accidents must generate injury or property damage.
- Accidents occur when random variables interact.
- Accidents can result from an act of God or nature.
- Accident investigations must determine fault.

them to take risks during job or task accomplishment. Heinrich believed that removing a "single" causal factor from a potential situation could result in preventing an accident. Interrupting or breaking the "accident cycle" by preventing unsafe acts or correcting an unsafe condition could reduce accident risk for individuals engaging in risky behaviors. Heinrich proposed a "five-factor" accident sequence in which a single causal factor could actuate the next step in the accident-cycle process. The five accident factors that included a person's background and social environment precipitated engagement in faulty behavior. That behavior interacted with a workplace mechanical or physical hazard causing an accident to occur. The accident occurrence then resulted in personal injury or property damage. Heinrich's conclusions pointed to what we now refer to as multiple causation theory.

System safety methods assume that accidents and mishaps result from multiple causal factors. System "thinking" views hazards and causal factors as moving in logical sequences to produce accident events. Refer to Chapter 4 for additional information on system safety methods.

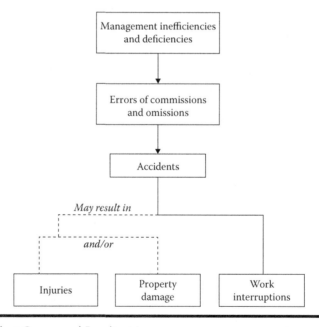

Sequence of Accident Causes and Resultant Losses

Traditional approaches to accident prevention simply classify causal factors as "unsafe acts" and "unsafe conditions." Hazard control personnel should use root-cause processes to discover, document, and analyze accident causal factors. Accident investigations and root-cause analyses (RCA) should focus on discovering information about system operation, deterioration, and original design errors. System-related hazard control efforts focus on unsafe system conditions and the interaction of human factors with these and other hazards. When a hazard contacts or comes close to another hazard, the result can cause death, injury, or property damage. Hazard closing occurs when two or more hazards attempt to occupy the same space at the same point in time. Hazard closing could be referred to as accident generation, with the result being an accident or mishap. These events can result in property damage, personal injuries, or both. Hazard closings can also result in close calls, near hits, or near-miss events. Hazard control management recognizes and acknowledges that an accident event occurs at a specific point in time. Many times anticipated or previously identified causal factors interact and create a mishap. These uncontrolled primary factors can set the accident generation cycle into motion. Hazard control efforts must eliminate the hazard or dangerous situation to reduce or eliminate the potential for harm. System thinking promotes the concept of providing separation between an individual and potential operational hazards. The hazard may remain within the system but in a controlled state. Attempt to reduce hazardous exposures by providing controls such as warning systems, monitoring equipment, and danger information. Attempt to motivate safe behavior through education, training, and supervision.

Multiple-Causation Theory

This theory promotes the idea that accidents result from various hazards or other factors interacting in some manner. Accident prevention professionals use different terms to describe these factors. Some refer to the factors as primary and secondary causes, whereas others use terms such as *immediate and contributing causes, surface and root causes*, or *causes and subcauses*. Most investigators agree that accidents happen due to multiple and sometimes complex causal factors.

Causal factors seldom contribute equally in their ability to trigger an event or contribute to accident severity. Accidents result from some type of interaction of causal factors. Human factors such as an unsafe act, error, poor judgment, lack of knowledge, and mental impairment can interact with other contributing factors, creating an opportunity for an accident to occur.

Biased-Liability Theory

Biased liability promotes the view that once an individual becomes involved in an accident, the chances of that same person becoming involved in a future accident increases or decreases when compared to other people. The accident-proneness theory promotes the notion that some individuals will simply experience more accidents than others because of some personal tendency.

Accident Pyramid

Heinrich introduced the "accident pyramid" in his book *Industrial Accident Prevention: A Scientific Approach*. This pyramid showed his accident-causation theory. Heinrich believed that unsafe acts led first to minor injuries, and then, over a period of time, to a major injury event. The accident pyramid proposed that 300 unsafe acts produced 29 minor injuries and one major injury. The concept of the accident pyramid remained unchallenged many years. However, some recent studies challenge the assumed shape of the equilateral triangle used by Heinrich. Some professionals now believe the actual shape of the model would depend on organizational structure and culture.

Prevention of Fatal Events

The March 2003 edition of the *Journal of Professional Safety* contained an article titled "Severe Injury Potential." The article, authored by Fred Manuele, suggested that accident-prevention efforts should focus more on preventing fatal events. He highlighted some specific examples that lead to fatalities in industrial settings. His list included not controlling hazardous energy, no written procedures for hazardous processes, failing to ensure physical safeguards, using unsafe practices for convenience (risk perceived as insignificant), and operating mobile equipment in an unsafe manner.

Human Factors

Hazard control personnel and top management must recognize that human behavior can help prevent and cause accidents. Understanding human behaviors can pose a challenge to most individuals. Behavior-based accident-prevention efforts must focus on how to get people to work safely. Hazard control managers, organizational leaders, and frontline supervisors deal with human behaviors on a daily basis. The definition of behavior-based hazard control could read as follows: "the application of human behavior principles in the workplace to improve hazard control and accident prevention." Organizational issues such as structure and culture along with personal considerations impact individual behavior (Table 3.2).

The term *human factors* covers a wide range of elements related to the interaction between individuals and their working environment. Management styles, the nature of work processes, hazards encountered, organizational structure, cultures, and education or training can influence individual behaviors. Hazard control efforts must address factors that impact human conduct or behavior. Organizations must ensure all members possess the knowledge, skill, and opportunity to act in preventing accidents. Motivation to act responsibly relates to a person's desire to do the right thing. Human factors can refer to personal goals, values, and beliefs, which can impact an organization's expectations, goals, and objectives. Flawed decision-making practices and poor work practices can create atmospheres conducive to human error.

Causal links between the accidents and unseen organizational factors may not appear obvious to most investigators. However, these factors could easily interact with trigger mechanisms to contribute to accident events.

Error

An essential component of accident prevention relates directly to understanding the nature, timing, and causes of errors. Error, as a normal part of human behavior, many times becomes overlooked during accident investigations and analysis processes. Errors can result from attention failure, a memory lapse, poor judgment, and faulty reasoning. These types of errors signify a breakdown in an individual's information-processing functions. When analyzing accident causes, hazard control personnel may never determine individual intent. However, focusing on

Table 3.2 Understanding Human Issues

- Character is the moral and/or ethical structure of individuals or groups.
- Belief refers to a mental act and habit of placing trust in someone or something.
- Value is based on what is believed and things that have relative worth or importance.
- Culture can be defined as socially accepted behaviors, beliefs, and traditions of a group.
- Attitude is the state of mind or feeling about something that is many times hidden.
- Behavior is the open manifestation of a person's actions in a given situation.

Table 3.3 Common Unsafe Acts

- Not following established job procedures
- Cleaning or repairing equipment with energy hazards not locked out
- Failing to use prescribed PPE
- Failing to wear appropriate personal clothing
- Improperly using equipment
- Removing or bypassing safety devices
- Operating equipment incorrectly
- Purposely working at unsafe rates or speeds
- Accomplishing tasks using incorrect body positions or postures
- Incorrectly mixing or combining chemicals and other hazardous materials
- Knowingly using unsafe tools or equipment
- Working under the influence of drugs or alcohol

the nature of behaviors in play at the time of error occurrence can provide some insight. When categorizing errors, attempt to differentiate between those occurring during accomplishing skilled behavior tasks and those related to unskilled tasks such as problem solving. Rule-based errors can occur when behaviors require the application of certain requirements. Process errors occur when individuals lack an understanding of procedures or complex systems. Knowledge-based errors occur when an individual lacks the skill or education to accomplish a certain task correctly. Planning errors occur when individuals fail to use a proper plan to accomplish a task. Finally, execution errors occur when using a correct plan but an individual fails to execute the plan as required (Table 3.3).

Motivating People

Motivating individuals to work safely requires the use of various approaches, depending on the situation. Little evidence exists that supports using punitive measures to motivate safe behaviors. The use of good human relations and effective communication skills can help improve individual motivation. The development of policies, procedures, and rules can never completely address all unsafe behaviors. Taking risks depends on individual perceptions when weighing potential benefits against possible losses. Some professionals believe that providing incentives to work safely, coupled with appropriate feedback, can enhance hazard control efforts in at least the short term. However, incentive programs with tangible rewards can also cause some individuals not to report accidents and injuries.

Accident-Deviation Models

Accident and mishap "deviation models" used in system safety processes can permit analysis of events in terms of deviations. The value assigned to a system "variable" becomes a deviation whenever it falls outside an established norm. When measuring system variables, these deviations can assume different values, depending on the situation. Hazard control policies and procedures

Table 3.4 Promoting Employee Involvement

- Develop an open-door policy to provide employee access to managers.
- Develop an easy-to-use accident, injury, and hazard reporting system.
- Require supervisors to conduct periodic safety meetings.
- Develop off-the-job safety and health education objectives.
- Encourage employee participation in job-hazard analyses.
- Disseminate hazard control and accident information in a timely fashion.
- Place more emphasis on people and less emphasis on compliance.
- Solicit employee suggestions on using hazard control resources and funds.
- Mandate hourly employee representation on the hazard control committees.
- Conduct special shift-worker education and training sessions.
- Conduct periodic organizational perception surveys about hazard control efforts.

should detail any specified requirements. A deviation from a specified requirement could result in a human error for failure to follow procedures. Therefore, we must consider incidental factors as deviations from an accepted practice.

An "unsafe act" relates to a personal action that violates or deviates from a commonly accepted safe procedure. Time functions as the basic dimension in a system deviation model, since incident analysis becomes a linear process rather than focusing on a single incident or a series of causal factors. Consider appropriate prevention measures by focusing on prior assessments and evaluations of the entire system. Active protection control would require constant repetitive actions on an individual. Passive protection controls would use relevant automatic protection innovations. Active interventions would require modifying and sustaining behavioral changes. Always stress the importance of behavior change or modification rather than any need for additional education. An effective approach to behavior modification may also require actions such as redesigning equipment or modifying physical environments to achieve the practice of safe behaviors.

When top management shows concern and commitment to the hazard control, individuals will more likely support organizational initiatives. The accountability factor for individual performance can serve as a key motivator for many individuals. Top management should continually remind everyone that the organization views hazard control as a priority function. Recognize and reward individuals for supporting hazard control efforts. Promote the use of tool-box talks, personal coaching, and supervisory involvement to promote hazard control (Table 3.4).

Accident Reporting

The timely and accurate reporting of accidents and injuries permits an organization to collect and analyze loss-related information. This information can help determine patterns and trends of injuries and illnesses. Organizations should encourage reporting by all members. The reporting process should focus on the importance of tracking hazards, accidents, and injuries, including any organizational trends. Educate all personnel to understand the need for maintaining a systematic process that accurately and consistently provides updated information. The system must not only permit data collection but also provide for a means to display any measure of success or failure in resolving identified hazards. Maintain records that enable managers at all levels to access data. Information made available to managers can assist them with changing policies, modifying operational procedures, and providing job-related training. Senior leaders must ensure the use of a system that meets the needs of the organization. Many vendors now offer accident/injury reporting and tracking software. Many of these processes permit the creation and printing of electronically generated reports. Using electronic reporting can help the organization save time and money. Make all the forms accessible on your internal computer network. Electronic reporting will encourage people to conform to your expectations and report near misses, accidents, and injuries in a timely manner. A sample of a completed form will help show and remind users what information you are seeking from them. Include a phone number and e-mail address for the person who can answer questions as they arise. Provide instructions for where and how to submit the report once it is completed. Electronic submission will save paper and retyping or scanning.

Input and track all safety incidents across your organization through one centralized online portal accessible to all employees and locations. New technology lets users easily create their own online forms. Customize fields and drop-downmenus to build an incident reporting form unique to your organization's needs. They can create incident reports sorted by employee, work group, unit, or department on type of injury or body part. Report incident frequency and severity rates and easily disseminate using an appropriate electronic format.

Customizable dashboards provide instant access to safety incident metrics from across your organization, providing real-time data and analysis needed to drive continual improvement. Manage the entire accident life cycle of incident reporting, responding, investigating, taking corrective action, tracking, and developing summary reports. Incident-reporting forms can trigger automatic, escalating follow-up e-mails to employees responsible for corrective actions, ensuring prompt resolution. Technology makes reporting and analysis easier and quicker than ever before. Organizations can no longer make excuses for not accurately collecting, tracking, and evaluating accident, hazard, and injury information.

Accident Investigations

A successful accident investigation must first determine what happened. This leads to discovering how and why an accident occurred. Most accident investigations involve discovering and analyzing causal factors. Conduct accident investigations with organizational improvement as the key objective. Many times an investigation focuses on determining some level of fault. Some organizations seek to understand the event to prevent similar occurrences. Most large organizations would benefit if the responsible supervisor conducted the initial investigation. Accident causal factors can vary in terms of importance, because applying a value to them remains a very difficult challenge. Sometimes investigators can overvalue causal factors immediately documented after the accident. These "immediate causal" factors reveal details about the situation at the time of occurrence. Investigators can also devalue less obvious causes that remained removed by time and location from the accident scene.

Identifying and understanding causal factors must include a strong focus on how human behaviors contributed to the accident. Organizations must conduct comprehensive investigations when preliminary information reveals inconsistencies with written policies and procedures. Using sound investigational techniques can help determine all major causal factors. Investigations should seek to find out what, when, where, who, how, and, most of all, why. Organizations must work to reduce the time between the reporting of a serious accident and the start of the investigation. Provide investigators with all the necessary tools and equipment to conduct a thorough investigation. Investigators must also seek to identify any failures of organizational-related management systems. These managerial failures contribute directly or indirectly to many accident events. Before attempting to discover and document causal factors, examine the site, and take steps to preserve any important evidence. Attempt to identify and document a list of witnesses. In some investigations, a particular physical or chemical law, an engineering principle, or equipment operation may explain a sequence of events. Some investigators use diagrams, sketches, and measurements to support their understanding of what happened. This type of information can also assist in future analysis efforts.

Consider using video recorders or cameras when conducting investigations. Use charts or sequence diagrams to help develop a probable sequence of events. Sequence charts can also help to understand events that occurred simultaneously. When conducting an investigation, attempt to look beyond the obvious to uncover direct and indirect causal factors. Evaluate all known human, situational, and environmental factors. Research studies indicate that most workplace accident investigations reveal 10 or more causal factors. Determine what broken equipment, debris, and samples of materials need removal from the scene for further analysis. Investigators should develop written notes about any items removed from the scene, including their positions at the accident scene. Make every effort to interview as many witnesses as possible before leaving the scene. These witnesses can serve as primary sources of information in many investigations. Accident-scene tampering and evidence removal before the investigator arriving would make witnesses very important.

Classifying Causal Factors

Understand the importance of initially documenting and classifying causal factors in one of the following three categories. The first category relates to operational factors such as unsafe job processes, inadequate task supervision, lack of job training, and work-area hazards. The second category relates to human motivational factors, including risky behaviors, job-related stress, poor attitudes, drug use, and horseplay. The third category relates to organizational factors, including inadequate hazard control policies and procedures, management deficiencies, poor organizational structure, or lack of senior leadership.

Interviewing Witnesses

Interview witnesses individually and never in a group setting. If possible, interview a witness at the scene of the accident. It also may be preferable to carry out interviews in a quiet location. Seek to establish a rapport with the witness, and document information using their words to describe the event. Put the witness at ease, and emphasize the reason for the investigation. Let the witness talk, and listen carefully and validate all statements.

Take notes or get approval to record the interview. Never intimidate, interrupt, or prompt the witness. Use probing questions that require witnesses to provide detailed answers. Never use leading questions. Ensure that logic and not emotion directs the interview process. Always close the interview on a positive note.

Accident Analysis

Organizations can use a variety of processes to analyze accident causal factors. Hazard evaluations and accident-trend analysis can help improve the effectiveness of established hazard controls. Routine analysis efforts can also enable organizations to develop and implement appropriate controls in work procedures, hazardous processes, and unsafe operations. Analysis processes rely on information collected from hazard surveys, inspections, hazard reports, and accident investigations. This analysis process can provide a snapshot of hazard information. Effective analysis can then take the "snapshots" and create viable pictures of hazards and accident causal factors.

When attempting to understand accident causes, hazard control personnel must identify, catalog, and then analyze the many factors contributing to an adverse event. Analyze to determine how and why an accident occurred. Use findings to develop and implement the appropriate controls. Do not overlook information sources such as technical data sheets, hazard-control committee minutes, inspection reports, company policies, maintenance reports, past accident reports, formalized safe work procedures, and training reports. When using accident-investigation evidence, remember the information can be in physical or documentary form. It can come from eyewitness accounts or from documentary evidence. The analysis must evaluate sequence of events, extent of damage, human injuries, surface causal factors, hazardous chemical agents, sources of energy, and unsafe behaviors. Consider factors such as horseplay, inadequate training, supervisory ineffectiveness, weak self-inspection processes, poor environmental conditions, and management deficiencies. A good accident analysis should create a word picture of the entire event. Refer to photos, charts, graphs, and any other information to better present the complete accident picture. The final analysis report should include detailed recommendations for controlling hazards discovered during the investigation and analysis.

Root-Cause Analysis (RCA)

RCA processes can help connect the dots of accident causation by painting a picture that includes underlying or beneath-the-surface-causes. Organizations many times fail to use effective and systematic techniques to identify and correct system root causes. Best-guess corrective actions do not address the real causes of accidents. Ineffective quick-fix schemes do not change processes to prevent future incidents. RCA focuses on identifying causal factors and not placing blame. A root-cause process must involve teams using systematic and systemic methods. The focus must remain on the identification of problems and causal factors that fed or triggered the unwanted event. When analyzing a problem, we must understand what happened before discovering why it happened. Do not overlook causal factors related to procedures, training, quality processes, communications, safety, supervision, and management systems.

An effective RCA process lays a foundation for designing and implementing appropriate hazard controls. Root causes always preexist the later-discovered surface causes. When root causes go unchecked, surface causes will manifest in the form of an unwanted event. For reasons of simplicity, system-related root causes fall into two major classifications. The first class concerns design flaws such as inadequate or missing policies, plans, processes, or procedures that impact conditions and behaviors. The other category, known as operational weaknesses, refers to failures related to implementing or carrying out established policies, programs, plans, processes, or procedures. When discovered and validated, specific root causes can provide insight to an entire process or system. The process can also help identify what fed the problem that impacted the system. Finally, RCA provides insight for developing solutions or changes that will improve the organization (Table 3.5).

Accident Reports

When preparing an accident-investigation report, use the analysis results to make specific and constructive recommendations. Never make general recommendations just to save time and effort. Use previously drafted sequences of events to describe what happened. Photographs and diagrams may

Table 3.5   Common Causal Factors

Poor Supervision

- Lack of proper instructions
- Job and/or safety rules not enforced
- Inadequate personal protective equipment (PPE), incorrect tools, and improper equipment
- Poor planning, improper job procedures, and rushing the worker

Worker Job Practices

- Use of shortcuts and/or working too fast
- Incorrect use or failure to use protective equipment
- Horseplay or disregard of established safety rules
- Physical or mental impairment on the job
- Using improper body motion or technique

Unsafe Materials, Tools, and Equipment

- Ineffective machine guarding
- Defective materials and tools
- Improper or poor equipment design
- Using wrong tool or using tool improperly
- Poor preventive maintenance procedures

Unsafe Conditions

- Poor lighting or ventilation
- Crowded or poorly planned work areas
- Poor storage, piling, and housekeeping practices
- Lack of exit and egress routes
- Poor environmental conditions such as slippery floors

save many words of description. Identify clearly if evidence is based on facts, eyewitness accounts, or assumptions. State the reasons for any conclusions, and follow up with the recommendations. An accident-analysis process must consider all known and available information about an event. Clarify any previously reported information, and verify any data or facts uncovered during the investigation. Review and consider witness information and employee statements or suggestions.

Organizational Functions That Support Accident Prevention

Interfacing Support Functions

The noninvolvement of staff functions and service components in hazard control can hinder success. Do not ignore a natural interface of hazard control management with other support functions such as facility management, purchasing, and human-resource management. Virtually every department and function of a modern organization contributes in some way to the effectiveness of the hazard control management function. There should be a greater interrelationship among staff and support functions that interface with accident prevention such as personnel, procurement, and maintenance, and so on. Too often, these functions act like they are encased in parallel tracks with little or no interaction. They must work in harmony to make an impact on preventing accidents and controlling hazards.

Operational and Support Functions

Operational and line elements must be conscious of their roles in accident prevention in terms of organizational policy, regulations, procedures, safety inspections, and other activities to support the hazard control function. Frequently, technical advances result in acceleration of organizational activities without the provision for accompanying related safeguards by management. Planning, research, budget, and legal functions must interface with accident-prevention programs and hazard control efforts.

Human Resources

As far as practical, the human-resource function should recruit, evaluate, and place the right person in the right job in terms of physical ability and psychological adaptability. Incorporate into job descriptions specific physical requirements, known hazards, and special abilities required for optimum performance. Human-resource professionals must identify all hazardous occupations and determine the knowledge, skills, abilities, physical requirements, and medical standards required to perform the job in a safe manner. All employees must receive appropriate orientation, training, and education necessary to support safe job accomplishments.

Facility Management

Facility management functions should ensure proper design layout, lighting, heat, and ventilation in work areas. Review specifications for new facilities, major renovations to existing facilities, and any plans for renting or leasing new work or storage areas. Maintenance activities should provide preventive maintenance service to avoid breakdown of equipment and facilities. Coordinate efforts with engineering, purchasing, and safety in reporting obsolete and/or hazardous equipment. Ensure that maximum safety is built into the work environment. It is much more efficient to correct a hazardous situation than to guard it or instruct employees to avoid it. It costs more to have accidents than to prevent them (Table 3.6).

Table 3.6   Facility Management Hazard Control Issues

* Coordinate implementation of hazard controls during design of all work areas.
* Take appropriate engineering action to eliminate or guard known hazards.
* Develop regular and preventive maintenance schedules for all tools and equipment.
* Implement regular inspections to identify equipment- and material-related hazards.
* Take immediate action to eliminate identified hazards and unsafe conditions.
* Ensure the proper design of work areas and the layout of equipment.
* Require each work unit to maintain high housekeeping standards.
* Ensure the installation of proper lighting, ventilation, and environmental controls.

Purchasing/Receiving

The organizational purchasing, contracting, or material management functions must consider safety requirements and standards when ordering equipment, tools, and other supplies. With the exception of certain legal requirements, the purchasing function should never attempt to dictate the safety-related standards or specifications of equipment, machines, and supplies requisitioned by hazard control or other operational functions. Hazard control managers should inform purchasing personnel about necessary standards, requirements, and safety factors that should be included in specifications. In addition to getting safety features incorporated into specifications, it is essential that the receiving function perform a detailed inspection of the equipment upon delivery. Never accept items that do not meeting safety specifications or standards. Ensure the receipt of safety data sheets when accepting hazardous materials. Purchasing functions must follow up and provide information concerning hazardous supplies and equipment.

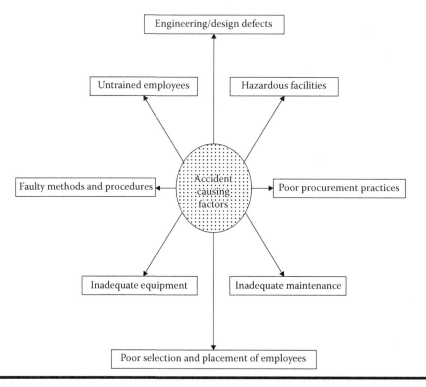

Inadequate Management Factors That Generate Accidents

Employee Health

Occupational health professionals and hazard control managers must coordinate and communicate issues on a continuous basis. Provide prompt emergency treatment of all injuries and illnesses. The coordination of the safety and employees' health functions can help ensure that workers get the best training, education, and protection from hazards. Recommend a multidisciplinary approach to manage health, risks, and costs. All work-related incidents of injury and exposure allegations should be reported immediately to employee health. Employee health should monitor, manage, or coordinate all workers' compensation injuries, reports on progress, imposition of

Table 3.7   Employee Health Program Components

- Blood-borne Exposure Control Plan (29 CFR 1910.1030)
- Fitness-for-Duty (Local Policies)
- Personal Protective Equipment (29 CFR 1910.132)
- Eye Protection (29 CFR 1910.133)
- Fire Safety (NFPA 101, 29 CFR 1910.38, Local and State Codes)
- TB Policy (CDC Guidelines and Health Department Requirements)
- Immunizations (CDC and Health Department Recommendations)
- Radiation Safety (29 CFR 1910.1096)
- Reproductive Hazards (OSHA, NRC, & NIOSH Recommendations)
- Confidentiality of Medical Records (HIPAA and OSHA Standards)
- Hazard Communication (29 CFR 1910.1200)
- Substance Abuse (Local Policies)
- Work-Related Injuries (29 CFR 1904) & Worker's Compensation Statutes
- OSHA Record Keeping (29 CFR 1904)
- Hearing Protection (29 CFR 1910.95)
- Work-Related Stress & Shift Work (NIOSH Publications)

necessary work restrictions, and return-to-work evaluations. Preemployment placement evaluations should focus on job-related issues with a thorough job analysis as part of the evaluation. If the evaluation indicates no medical causes for performance problems, refer the employee back to management for appropriate administrative action. A preplacement assessment develops a baseline for medical surveillance and helps determine capability of performing essential job functions.

The Americans with Disabilities Act (ADA) requires job descriptions for all job offers needing preplacement (post-job-offer) physical capacity determinations. Functional-capacity evaluations may help in determining job placement and modifications. Essential job functions can determine capability of a prospective employee to perform those functions with or without reasonable accommodations. Assessments may include an update of the occupational and medical histories, biological monitoring, and medical surveillance. Conduct a post-exposure assessment following any exposure incident. Determine the extent of exposure and develop measures to prevent recurrence.

Many organizations develop a formal Early Return to Work or Transitional Duty Program. Rehabilitation involves facilitating the employee's recovery to a preinjury or illness state. Occupational health should be informed about the rehabilitation of workers with any illnesses or injuries, including those not considered work-related. The goal of case management is to work with the employee to facilitate a complete and timely recovery (Table 3.7).

Shift Workers

Conduct mandatory education to inform individuals how to better cope with shift work. The human body follows a 24–25-hour period called the circadian clock. This internal clock regulates cycles in body temperature, hormones, heart rate, and other body functions. The desire to

Table 3.8 Shift Work Principles

• Shift differential pay alone does not improve worker morale or performance.
• Provide special orientation sessions for new shift workers.
• Never schedule organizational training after a work shift.
• Conduct job-training sessions before or during the scheduled work shift.
• Never schedule shift workers to attend training on their off days.
• Provide mandatory education sessions for all shift workers.
• Provide handouts (with the latest research) on how to improve sleep patterns.
• Encourage workers to share the information with their families.
• Provide flexible scheduling during a crisis and for special occasions.
• Encourage workers to communicate their feelings about the job with supervisors.
• Never promote overtime among shift workers.
• Short naps may be effective in improving worker alertness and productivity.
• Show concern about worker's off-the-job activities, such as traveling to and from work.
• Use professionals to help develop effective education programs.
• Encourage workers to seek medical assistance if needed.
• Encourage workers to report sleepiness when operating machines or equipment.

sleep is the strongest between the hours of midnight and 6:00 AM. This internal clock is difficult to reset, and up to 20% of night workers fall asleep on the job. Most sleep occurs during the second half of the shift. According to the National Sleep Foundation Poll, 65% of all people report that they do not get enough sleep. This translates into more health problems and impaired immune systems. The financial loss to business because of decreased productivity has been estimated at more than $18 billion each year. Workplace and vehicular accidents are higher for shift workers. Many shift workers hold down more than one job. Many times family members do not understand the need of those working nontraditional shifts. Second- and third-shift workers are more prone to stress-related problems than those working day shifts (Table 3.8).

Worker Compensation

Workers' Compensation laws ensure that employees injured or disabled on the job receive appropriate monetary benefits, eliminating the need for litigation. These laws also provide benefits for dependents of those workers who are killed because of work-related accidents or illnesses. Some laws also protect employers and fellow workers by limiting the amount an injured employee can recover from an employer and by eliminating the liability of coworkers in most accidents. State Workers' Compensation statutes establish this framework for most employment. Federal statutes are limited to federal employees or those workers employed in some significant aspect of interstate commerce. The injury or illness must have resulted from employment.

Workers' compensation provides benefits to the injured worker, including medical coverage and wages during periods of disability. Employer liability protects employers from litigation based on work-related injury. Employers can get coverage by purchasing commercial insurance, establishing a self-insurance program, attaining coverage in workers' compensation self-insurance or association funds, or being placed in a state-controlled fund. The state or the National Council of Compensation Insurance, an independent rating organization, normally determines basic rates paid by employers. The actual rates would be based on a number of other factors: (1) company or fund quoting the coverage, (2) classification code(s) of the employer, (3) payroll amount for the work force covered, and (4) experience rating.

The Federal Employment Compensation Act provides workers' compensation for nonmilitary, federal employees. Many of its provisions are typical of most worker-compensation laws. Awards are limited to "disability or death" sustained while in the performance of the employee's duties but not caused willfully by the employee or by intoxication. The act covers medical expenses due to the disability and may require the employee to undergo job retraining. A disabled employee receives two-thirds of his or her normal monthly salary during the disability.

The Long Shore and Harbor Workers' Compensation Act provides workers' compensation to specified employees of private maritime employers. The Office of Workers' Compensation Programs administers the act. The Black Lung Benefits Act provides compensation for miners suffering from "black lung" or pneumoconiosis. The Act requires liable mine operators to pay disability payments and establishes a fund administered by the Secretary of Labor providing disability payments to miners, where the mine operator is unknown or unable to pay. The Office of Workers' Compensation Programs regulates the administration of the act. The World Health Organization (WHO) defines impairment as any loss or abnormality of psychological, physiologic, or anatomic structures or functions.

The American Medical Association (AMA) defines impairment as loss, loss of use, or derangement of any body part, system, or function. WHO defines a disability as any restriction or lack of ability, resulting from an impairment, to perform an activity within the range considered normal. Most states require examiners to use the AMA "Guides to the Evaluation of Permanent Impairment" to determine accurate impairment ratings. The AMA guidelines limit the range of impairment values reported by different examiners.

Return-to-Work/Modified-Duty Positions

Establishing a realistic return-to-work function can save organization's financial losses due to fraudulent claims. Any "Return to Work Initiative" should accommodate injured workers by modifying jobs to meet their work capabilities. This action permits employees to become productive assets during their recovery. Early return-to-work programs can accelerate an employee's return by addressing the physical, emotional, attitudinal, and environmental factors that otherwise inhibit a prompt return. Senior management must commit to returning injured workers to productive roles. Implement performance measures to monitor the effectiveness of the program.

Develop profiles of jobs considered suitable for early-return participants. A profile should define the job in terms of overall physical demands, motions required, environmental conditions, the number of times the job is performed per week, and its duration. Conduct a systematic analysis of specific jobs for the purpose of modifying them to accommodate the unique needs of the injured worker. Individuals who are skilled in ergonomic task analysis, engineering, safety, and biomechanics should help perform the job analysis. Managed-care providers can also assist in job modifications. Communicate the availability of early-return jobs with care providers, claims adjusters, and the injured worker. Work with your managed-care provider and worker to move them to full production status in their assigned jobs as quickly as possible.

Substance Abuse

Substance use, misuse, abuse, and coping strategies can have a significant impact on mental health at work. Generally, substance use becomes a problem when an individual has lost control over their use and/or continues to use despite experiencing negative consequences. Employers should look for warning signs that indicate an employee may be struggling with substance abuse. Signs of substance abuse can appear similar to those caused by stress, lack of sleep, and physical or mental illness. Identify abuse by establishing preemployment, random, and "for cause" testing. Ensure the development and implementation of a testing policy.

Refer and evaluate employees addicted to performance-impairing drugs such as alcohol, narcotics, sedatives, or stimulants to qualified assistance or treatment facilities. Establish an agreement between the organization and the individual to address rehabilitation and random testing upon return to work. Measurable losses attributed to substance abuse can include absenteeism, overtime pay, tardiness, sick-leave abuse, health-insurance claims, and disability payments. Some of the hidden costs of substance abuse can include low morale, poor performance, equipment damage, diverted supervisory time, and low production quality. Losses can include legal claims, workers' compensation payments, disciplinary actions, security issues, and even dealing drugs in the workplace. Supervisors play the key role in maintaining an effective substance-abuse program (Table 3.9).

Orientation, Education, and Training

Orientation relates to the indoctrination of new employees into the organization. Orientation can be defined as the process that informs participants how to find their way within the organization. Usually safety and hazard control topics make up only a portion of any new-employee orientation session. Many well-meaning organizations attempt to present detailed safety and hazard control information during new-employee orientation sessions. However, attempting to provide too much performance-based safety education during orientation can be ineffective because of time constraints. Meeting the learning objectives must take precedence over simply documenting an educational session. New-employee orientation sessions must address the importance of safety, management's commitment, and worker responsibilities to practice good hazard control

Table 3.9   Signs of Substance Abuse

- Increased absenteeism
- Poor decisions and ineffectiveness on the job
- Poor quantity and/or quality in production
- High accident rates
- Resentment by coworkers who pick up the slack
- Poor morale in the department
- Late three times more often than other workers
- Uses three times more sick leave than others
- Five times more likely to file a worker's-compensation claim
- Involved in accidents four times more often than other employees

principles. Some performance-based Occupational Safety and Health Administration (OSHA) standards require that employees receive more detailed information than can be presented during an orientation session. This information should be presented in other education or training sessions.

Education is the incorporation of knowledge, skills, and attitudes into a person's behavior and includes the connotation of thinking. Education can provide information on topics previously trained. System safety methods use instructional system design (ISD) educational and training methods to ensure the competency of individuals working in or supporting operations. We refer loosely to this concept as simply adult education. ISD requires that required competencies be determined before educational and training sessions are developed. The sessions presented focus on the competencies both in the classroom and in realistic operational or job settings. Another example of ISD usage is in the construction industry, which developed "tool box" safety presentations many years ago to ensure workers practice safety on the job.

Sometimes, we use the phrases *job-related training* or *job safety training* to refer to instructional system design. Training relates to the acquisition of specific skills, whereas education refers to the incorporation of knowledge, skills, and attitudes into a person's behavior. Consider training as the process of presenting information and techniques that leads to competency of those participating. Conduct hands-on training outside of a classroom if possible, unless using realistic simulation processes. Effective training must strive to promote understanding, positively impact worker attitudes, and improve individual performance. Training must facilitate the transfer of knowledge and skills that relate to real-world activities. Many organizations do not dedicate sufficient time, allocate sufficient resources, or require attendance at training and education sessions.

Providing Adequate Sessions

An around-the-clock operation makes the education and training of shift employees even more challenging. Educate and train shift workers before or during their shift but never after the shift. Many organizations do not honestly evaluate training and education effectiveness. They maintain attendance or participation documentation. However, this documentation may not document and validate retention or competency. Some professional educators recommend documenting training and education attendance at the conclusion of the session. Suggest using a short quiz or performance assessment to document learning. Consider the use of employee safety meetings to educate workers about on- and off-the-job safety topics.

Publish an education and training policy statement to outline goals and objectives. Use various methods such as posters, flyers, bulletins, newsletters, self-study programs, classroom presentations, on-the-job training sessions, professional seminars, safety training fairs, and computer-assisted programs to communicate hazard control and safety topics. Some organizations delegate a number of training responsibilities to the individual departments. Other organizations use a full-time educational coordinator. Large or specialized departments in some organizations, such as laboratories, conduct most of their own training programs.

Training and Hazard Control

Hazard control managers and training personnel must coordinate education and training objectives to ensure they meet organizational needs. Conduct training for employees transferring to new jobs or work areas. Train those returning from an extended period away from the job and those new to the workforce. Schedule training sessions to match the needs of the organization and needs of learners. Always view education and training as organizational functions and not programs. When implementing an effective hazard control education and training function, consider the following elements: (1) identify needs, (2) develop objectives, (3) determine learning methods, (4) conduct the sessions, (5) evaluate effectiveness, and (6) take steps to improve the process.

Training must compliment and supplement other hazard controls and address rules and work practices. Some ways to evaluate training can include the following: (1) student opinions expressed on questionnaires, (2) conducting informal discussions to determine relevance and appropriateness of training, (3) supervisor's observations of individual performance both before and after the training, and (4) documenting reduced injury or accident rates. Revise the content of the session when an evaluation reveals that those attending did not show the knowledge or competency expected. Recurring sessions should cover on-the-job training and refresher sessions to ensure employees remain current in worker-related issues, including safety topics. Changes covered might include updated technology procedures, new government regulations, and improved practice standards. Engineering controls remain the preferred way of preventing accidents involving hazards related to unsafe mechanical and physical hazards. However, education and training serve as the most effective tools in preventing accidents by human causes. Through adequate instruction, people can learn to develop safe attitudes and work practices. Design education and training sessions by using clearly stated goals and objectives that reflect the knowledge and skill needs of people.

Training Methods

Instructional presentations can use a variety of methods to improve learning. The use of a pretest permits instructors to evaluate the knowledge of participants before the session begins. A pretest can motivate some participants to learn key concepts and principles. Informal discussions and lectures should incorporate time for questions and answers, to encourage participation. Demonstration methods permit the instructor to use a hands-on technique to promote the application of knowledge. Training content must directly apply to the hazards, procedures, equipment, and behaviors encountered on the job. People will receive instruction when they understand how they can apply the training to real-world situations. Since people learn in different ways, use a variety of training methods to promote learning. Some methods of education and training include lectures, videos, class discussions, demonstrations, written exercises, small group exercises, hands-on exercises, and combination methods. Some trainers develop and use games to review critical material, especially in refresher training sessions.

Consider ways to validate retention and learning. Methods often used include discussion, written tests and quizzes, trainee demonstrations or presentations, and on-the-job observation. Refer to OSHA booklet "Training Requirements in OSHA Standards and Training Guidelines" (OSHA 2254) for additional information about designing training sessions. Some situations permit students to meet training objectives by participating in a realistic scenario such as a disaster exercise. Many organizations use interactive software and other web-based learning opportunities to meet training objectives. Computer-generated and web-based sessions can permit the learner to control the flow of information during the training session.

Many organizations overlook the need to provide informational sheets and handouts to support training requirements and provide future reference information. Multimedia-presented visual aids in today's world can enhance learning. However, simply using computer-generated slides, overhead transparencies, white boards, videos, and flip charts do not guarantee the mastery of training objectives (Table 3.10).

Table 3.10 Basic Questions to Ask about Training Sessions

- Did the session cover critical issues or hazards?
- Did the presenter use an appropriate instructional method?
- Did the instructor cover all educational objectives clearly and concretely?
- Did the objectives state acceptable performance expectations of participants?
- Did the session simulate or address real situations?
- Did the participants show a motivation to learn during the session?
- Did the instructors encourage active participation by all participants?
- Did the presenter ask participants to critique or evaluate the session?

Off-the-Job Safety Education

Organizations should consider presenting off-the-job topics to all employees. Providing sessions that address off-the-job hazard control, safety, and health topics sends the message that the organization cares about its people. Present information addresses real-world issues such as summer and winter hazards, holiday safety, and traffic safety topics. Health topics could address eating healthy, the importance of exercise, and managing stress. Other topics to address could include home-related topics such as fire safety and fall prevention. Suggest presenting an off-the-job education on a monthly basis.

Instructional System Design (ISD)

Some organizations with complex systems or processes may benefit by implementing ISD educational and training methodologies. ISD can help organizations identify what an employee should know and what competencies related to hazard control he or she must show. The ISD approach promotes and supports acceptable performance by an employee or worker. ISD can also identify deficiencies in task knowledge and work competencies. Information and understanding about hazards and behaviors must be determined by system and job-tasks analysis. Understanding systems and processes can help hazard control managers validate knowledge and performance requirements of job or task. When preparing to develop an education and training plan on an unfamiliar procedure or system, the job-hazard analysis provides the foundation for success. Appropriate content can be added to the plan based on the following: (1) reviewing accident and injury records, (2) requesting workers to describe their job and related hazards, (3) observing and interacting with workers performing their job tasks, and (4) comparing content of other training plans dealing with similar hazards or risks.

Promoting Hazard Control

Seek ways to promote an interest in hazard control by helping organizational members develop safe work habits and by providing a hazard-free job environment. As addressed previously, make hazard control an organizational function and an operation priority. Organization size, type, and climate can impact hazard control promotional activities. Do not forget to include shift workers in any promotional campaign, because their support remains a crucial part of accident-prevention efforts. An effective promotional campaign shows management's commitment to hazard control. It also reminds employees to work safely and take ownership of their contributions. Recognition and other incentives can help promote hazard control importance, when designed and managed effectively. Provide employees the opportunity to participate in hazard control efforts. Encourage open communication and feedback among all organizational members. Continually stress the organization's commitment to a safe and healthy workplace.

Promote the use of workplace safety meetings to keep the lines of communication open and reinforce training concepts. The use of posters does not compensate for inadequate hazard control management, broken equipment, unsafe job procedures, poor supervision, or ineffective training. Posters, when used, should communicate a simple, straightforward message. Place in well-lighted locations such as lunchrooms, washrooms, entrances, and loading points. Many organizations rotate posters at regular intervals based on potential hazards.

A well-designed bulletin or newsletter can still effectively promote objectives of hazard control. An employee-oriented bulletin or newsletter should provide information and recognize accomplishments. A publication written sincerely in a straightforward manner will increase readership. Publications containing personalized articles can grab the reader's attention and prove very effective in promoting safety. Ask for suggestions, articles, feedback, and comments from readers. Develop some features or subjects that will appear on a regular basis. Look for ways to sell and promote the hazard control or safety message. Provide bulletin and newsletters in both print and electronic versions.

Many organizations fail to promote or require the use of regular work-site safety meetings. Safety meetings and tool-box talks keep the lines of communication open and reinforce training and education objectives. Supervisory safety meetings can promote safety and also encourage worker involvement in accident-prevention efforts. Meetings can motivate individuals to practice safety on the job. They also provide opportunities for employees to make suggestions, help pinpoint problem areas, and recommend corrective solutions for workplace hazards. Safety meetings provide supervisors an opportunity to address new procedures and equipment acquisitions.

Review Exercises

1. Define an accident in your own words.
2. List three common accident myths.
3. In your opinion, what was the central concept of Heinrich's Five-Factor Accident Sequence?
4. Explain the basic premise of any multiple-causation accident theory.
5. In your opinion, should accident-prevention efforts focus more on potential fatal events and less on other hazards?
6. What role do human-behavior principles play in effective hazard control?
7. Define an execution error.
8. List at least seven "common" unsafe human actions in the workplace.
9. Why would good human relations and communication skills help motivate people to act safely?
10. Explain the basic premise of accident "deviation" models.
11. List at least five ways that organizational leaders can promote employee involvement in hazard control efforts.

12. Define in your own words the following basic causal factor categories:
    a. Operational factors
    b. Motivational factors
    c. Organizational factors
13. When analyzing accidents, why would creating an accurate "word picture" be important to future prevention efforts?
14. Team RCA should focus on which two fundamental methods to ensure success?
15. Why must hazard control managers understand organizational structures and interfacing functions?
16. In your own words, describe the roles that human resources and purchasing play in supporting organizational accident prevention efforts.
17. How do "early return to work" initiatives and case-management activities support the employee health function?
18. List at least seven key concepts or principles related to shift work that management should consider implementing to improve accident-prevention efforts.
19. List at least five signs of employee substance abuse.
20. Explain the difference between education and training.

Chapter 4

Hazard Control—Related Disciplines

Introduction

The basic concepts and principles addressed in the previous chapters provided a foundation for implementing effective hazard control functions for a variety of organizational settings. This chapter provides a brief overview of selected related disciplines and emerging hazards. The chapter also provides an overview of common organizational hazards found in many organizations. The practice, art, and science of safety relates to actions taken to protect people, property, and the environment. This protection can impact physical, social, spiritual, financial, political, emotional, occupational, or psychological consequences of failure or loss. Hazard control management views safety as collective actions, innovations, and controls needed to prevent such losses. We must relate hazard control efforts to important organizational concepts such as quality, reliability, and maintainability. These principles tend to help determine the value assigned to the function of hazard control and accident prevention. Management deficiencies and inefficiencies increase organizational costs and can negatively impact organizational effectiveness. Hazard control helps prevent accidents and minimize organizational losses. It must never "just" focus on activities but promote actions that improve accident prevention and safety. A system approach to hazard control must encourage both operational and support personnel to better understand their roles in preventing losses. Viewing accidents from a system point of view can help organizations to quickly identify hazards and implement necessary controls. Hazard control managers must learn to use system techniques and methods to improve organizational performance.

Common Facility Hazards

Organizations must emphasize hazard control as a function of every unit, department, or division. Often overlooked, administrative areas contain a variety of hazards. These hazards can include material lifting, repetitive motions, tripping, and risks of electrical shock.

Make work area design a priority, and use ergonomic principles to help minimize work-related complaints, illnesses, or injuries. Some basic considerations would include evaluating force, duration, position, frequency, and metabolic expenditure of administrative workers. The most significant factor in the ergonomic equation addresses the working position of each individual. For example, administrative personnel working with computers for more than 4 hours daily can develop hand, arm, shoulder, neck, or back maladies (Table 4.1).

All personnel working at computer stations should frequently take short breaks to permit the eyes to relax. Glancing at an object at least 20 ft away can be of great help. Some workers can get relief by blinking or shutting their eyes for a few seconds at a time. Design the workstation with a padded keyboard, adjustable table, and tilting screen. Encourage workers to experiment and find a comfortable position. Design workstation to minimize glare and alleviate eyestrain. Remember, effectiveness of chair seating can vary with individual worker characteristics. Do not make the mistake of providing the same chairs to everyone. Chair backrests should support the entire back but also permit adjustments by the worker. Encourage workers to report potential ergonomic-related issues including pain, tingling, numbness, swelling, or other body discomforts.

Preventing Slips, Trips, and Falls

Organizations should take actions to reduce slip, trip, and fall hazards. Understanding the causal factors related to slip, trip, and fall events can help prevent them. The friction between the foot and shoe sole directly relates to a "relative force" known as the static coefficient of friction. The Americans with Disabilities Act (ADA) recommends a minimum of 0.6 on level walking surfaces and 0.8 on ramps. Occupational Safety and Health Administration (OSHA) requires a minimum of 0.5 on all walking surfaces. Organizations must accept the obligation to conform to "standard of care" requirements for flooring selection and maintenance. Many facilities and organizations simply fail to take appropriate actions to ensure conformance to a reasonable standard of care. Failure to do so can result in "proximate cause" or "contributory negligence" incidents. Organizations should install appropriate flooring for the location, ensure proper maintenance, and provide appropriate cleaning for all walking locations (Table 4.2).

Electrical Safety

Hazard control personnel must ensure assessment of all work areas to identify unsafe electrical hazards. The pressure of current traveling through electrical conductors is known as volts. Resistance relates to the measured flow of electricity. We describe the measurement using the term *ohms*. Some materials such as metal offer little resistance and become conductors very easily. Other substances,

Table 4.1 Computer Work Area Evaluation Considerations

- Layout and design of each work area and nature of tasks performed.
- Ergonomic hazards present, including position and repetition.
- Evaluate degree of postural constraints of the workers.
- Assess the impact of the pace of work required for task accomplishment.
- Establish and require appropriate rest schedules or task breaks.
- Determine personal attributes of individual workers.

Table 4.2   Safe Flooring Suggestions

- When selecting flooring, consider performance factors in wet and dry conditions.
- Ensure selected floor will remain durable in high-traffic areas.
- Consider "abrasion resistance" or how long a surface will retain its slip resistance.
- Establish maintenance and care procedures recommended by manufacturer.
- Evaluate the flooring's impact resistance in areas with heavy loads.
- Consider safety as more important than appearance.
- Evaluate the life-cycle costs and other expenses beyond cost of flooring.

Table 4.3   Shock Severity Factors

- Amount of current (amperes) flowing through the body
- Path the current takes through the body
- Length of time the person remains in the circuit
- Phase of the heart cycle when shock occurs
- General health of the person involved

such as porcelain and dry wood, offer high resistance. Materials that prevent the flow of electricity are called insulators. Electricity travels in closed circuits and routes itself through a conductor. Electrical shock occurs when the body becomes part of that circuit. Shock normally can occur when a person contacts both wires of an electrical circuit, contacts one wire of an energized circuit and the ground, or contacts a "hot" or energized metallic part while in contact with the ground.

Severe shock can cause falls, cuts, burns, and broken bones. Three types of burns can result from shocks: (1) electrical burns result from current flowing through tissue or bone—the damage is from the intense heat that damages tissues; (2) thermal burns occur when the skin comes into contact with the hot surfaces of overheated conductors or other energized parts; and (3) arc burns are caused by high temperatures near the body and are produced by an electrical arc or explosion (Table 4.3).

The use of circuit-protection devices can limit or shut off the flow of electricity in the event of a ground-fault overload or a short circuit in the wiring system. Fuses and circuit breakers function as over-current devices that automatically open or break when the amount of current becomes excessive. Fuses and circuit breakers primarily protect equipment and conductors. Ground-fault circuit interrupters (GFIs) devices shut off electrical power immediately by comparing the amount of current going to the equipment and the amount returning along the circuit conductors. Use GFI devices in all wet locations and construction areas to protect people from shock hazards. Ensure all electrical installations and related equipment adheres to requirements found in NFPA/ANSI 70, National Electric Code® (NEC). This code applies to every replacement, installation, or use of any electrical equipment. Refer to 29 CFR 1910.331–335 and NFPA 70E for electrical safety-related work practice requirements (Table 4.4).

Noise Hazards

OSHA requires that organizations with workers exposed to decibel levels exceeding 85 decibels (dBA) implement a Hearing Conservation Program. OSHA identifies 90 dBA, based on the results of an 8-hour time-weighted average (TWA), as the safe level of noise exposure. The 90 dBA

Table 4.4 Electrical Codes and Standards

- NFPA 70B, Recommended Practice for Electrical Equipment Maintenance
- NFPA 70E, Electrical Safety Requirements
- NFPA 72, Fire Alarms
- NFPA 110, Emergency and Standby Power Systems
- NFPA 111, Stored Electrical Energy Emergency and Standby Power Systems
- 29 CFR 1910.147, Control of Hazardous Energy (Lockout/Tagout Requirements)
- 29 CFR 1910 Subpart S, Electrical/Safeguarding Employees

concentration refers to the OSHA permissible exposure limit (PEL) for workplace noise exposure. An 8 hour TWA exceeding 90 dBA requires the employer to implement control measures to reduce the exposure to 90 dBA or below. You measure frequency or pitch cycles per second using the term *hertz* (Hz). We express the measurement of amplitude or intensity as decibels (dB). A decibel scale uses a logarithmic measure with the increase of each 10 dB perceived as being twice as loud. However, perceived loudness remains a subjective expression. Instruments used to monitor noise levels include sound-level meters and noise dosimeters. Both devices measure noise using decibel readings. A decibel unit expresses a logarithmic ratio to an established reference level. Sound-level meters and noise dosimeters usually measure on two or three different frequency scales. Frequency refers to the number of vibrations per second a noise contains. Meters measure in hertz and use frequency scales of A, B, or C. OSHA requires taking noise measurements using the A scale. Sound meters provide only a one-time measurement of sound levels. Use sound meters to determine which areas require a dosimeter measurement. OSHA also requires employers to evaluate sound attenuation provided by ear protectors, considering the environment in which they are used.

The Environmental Protection Agency (EPA) developed the noise-reduction rating (NRR) as a method of gauging the adequacy of a hearing protector's noise-reducing capacity. The NRR equates to the amount of decibels by which a device reduces noise exposure. When exposed to a TWA of 100 dB and the worker wears a pair of earmuffs with a NRR of 26, the calculated actual exposure would be 74 dB.

Conduct training and education annually for all employees. Training must cover the effects of noise on hearing and the purposes of using hearing protectors. Training must also address the advantages, disadvantages, and attenuation of various hearing protectors. Employees must know how to select, fit, use, and maintain protectors. Conduct a baseline audiogram within 6 months after confirming exposure at or above 85 dB action level. Conduct the initial annual audiogram within one year of baseline. Employers must provide testing free of cost to employees. Ensure the calibration of audiometers meet ANSI standards.

Heating, Ventilation, and Air Conditioning (HVAC) Systems

An HVAC system includes all heating, cooling, and ventilation equipment serving a building. HVAC systems can include furnaces or boilers, chillers, cooling towers, air-handling units, exhaust fans, duct work, filters, steam, or piping. A good ventilation design should facilitate the uniform distribution of "supply" air. Install exhaust fans a significant distance away from supply

Table 4.5   Key ASHRAE Ventilation Standards

- ASHRAE 52, Testing Air-Cleaning Devices for Removing Particulates
- ASHRAE 55, Thermal Environmental Conditions for Human Occupancy
- ASHRAE 62, Ventilation for Acceptable Air Quality

sources or intake vents. The American Society of Heating, Refrigeration, and Air Conditioning Engineers (ASHRAE) publish consensus standards and other guidelines addressing installation and maintenance requirements. Local exhaust ventilation systems must conform to the construction, installation, and maintenance requirements found in ANSI Standard, Fundamentals Governing the Design and Operation of Local Exhaust Systems, Z9.2 and ANSI Z33.1. OSHA Standard 29 CFR 1910.107 addresses local exhaust duct systems, independent exhaust, and room intakes. OSHA Standard 29 CFR 1910.106 addresses requirements for inside storage rooms with flammable materials. OSHA also publishes tables and standards for Air Contaminants in 29 CFR Subpart Z. OSHA standards cover requirements for a variety of operations, including abrasive blasting, grinding/polishing operations, and spray finishing (Table 4.5).

Indoor Air Quality

Maintaining indoor air quality requires trained professionals who understand the entire building and its systems. Cleaning improves the quality of all indoor environments. Good cleaning practices can also reduce human concern and anxiety. We must always consider ventilation and adequate air flow within buildings as a key element of any cleaning process. Always clean buildings for safety and health first. Sick building syndrome (SBS) refers to buildings that may cause human illness, but positive confirmation remains elusive. Building-related illness refers to an illness related to some type of exposure from a built structure. Some type of testing and analysis confirms the illness came from the structure. Legionnaires' disease remains a common example in the minds of many. Multiple chemical sensitivity (MCS) relates to a person's exposure to many chemical or hazardous substances in small concentrations that causes health-related problems. Organizations should focus on proper building care, HVAC maintenance, and intelligently designed renovations to reduce indoor air quality hazards. Analyze ventilation systems to ensure comfort, ventilation, and sanitation. Inspect accessible areas for obvious problems, poor design, or signs of contamination. Determine airflow, temperature, humidity, carbon dioxide concentration, and air pressure differentials. Conduct periodic HVAC inspections to ensure an adequate supply of outside air. For additional guidance refer to NIOSH/EPA Publication: Building Air Quality—A Guide for Building Owners and Facility Managers (Table 4.6).

General Machine and Tool Safety

Educate personnel about the applications, limitations, operations, and hazards related to their tasks, tools, and equipment. Provide sufficient racks, shelves, or toolboxes for storing tools. Require personnel working on ladders, scaffolds, or platforms to carry required tools in a proper bag. When working with tools that create hazards, require workers to wear appropriate personal protective equipment (PPE). Identify all tasks requiring the use of safety glasses or goggles. Do not

Table 4.6 Mold Prevention Tips

- Repair water and plumbing leaks.
- Fix the source of any moisture or water incursion as soon as possible.
- Prevent moisture from condensing by increasing surface temperatures.
- Insulate or increase air circulation.
- Keep HVAC drip pans clean, flowing properly, and unobstructed.
- Perform scheduled building/HVAC inspections and maintenance.
- Maintain indoor relative humidity below 70% (25%–60% is preferred).
- Vent moisture-generating appliances to the outside if possible.
- Clean/dry damp spots within 48 hours after discovery.
- Provide good drainage, and slope the ground away from foundations.

permit electric power tools to be used in close proximity to flammable vapors, dusts, construction materials, or wet environments. Ensure all electric tools are properly grounded or equipped with a GFI. Cover or guard power-transmission exposures such as belts, pulleys, shaft, or gears. Grinder operators must never grind on the side of the wheel, unless designed for that purpose. Never force or jam work onto the wheel. Attach all guards securely, and ensure that the face of the wheel does not contain foreign materials, chips, or nicks. Evaluate grinders producing excessive vibrations. Secure adjustable tongues at least 1/4 inch or less from the wheel. Install work rest tolerances at least 1/8 inch from the wheel. Secure the work rest at the center of the wheel or above.

Powered Industrial Trucks (29 CFR 1910.178)

The OSHA Standard incorporates by reference many OSHA standards as well as general consensus standards. Areas addressed include description of truck design, approval, and labeling requirements. Preventing accidents requires comprehensive worker training, systematic management, safe environments, safe trucks, and proper work practices. National Institute for Occupational Safety and Health (NIOSH) investigations of forklift-related deaths indicate that many workers and employers remain unaware of many risks and hazards.

Many operators simply fail to follow published procedures. Industrial trucks and forklifts must bear a label or some other identifying mark indicating approval by the testing laboratory as required by OSHA and ANSI standards. The storage and handling of liquid fuels must adhere to requirements of NFPA 30. Concentration levels of carbon monoxide gas created by powered industrial truck operations must not exceed the levels specified in 29 CFR 1910.1000. Dock boards or bridge plates must meet the requirements of 29 CFR 1910.30. OSHA requires employers to develop a complete training program. OSHA mandates training of operators of powered industrial trucks before operating them on the job. Training must include classroom-type and practical hands-on training. Focus on proper vehicle operation, the hazards of operating the vehicle in the workplace, and the requirements of the OSHA standard. Operators must go through an evaluation at least once every 3 years to validate that their skills remain at a high level. Conduct refresher training whenever needed. Employers must certify completion of all training and evaluations.

Ladder and Scaffold Safety

OSHA addresses ladder safety in both their General Industry and Construction Standards. The American National Standards Institute (ANSI) also publishes ladder consensus standards. Individuals working at elevations of more than 4 ft encounter a serious risk of injury because of falls. Most ladder-related injuries result from unsafe or improper use. Train workers never to use ladders with broken rails, rungs, or missing hardware. Never permit personnel to use portable metal ladders when performing work on or near electrical equipment. Emphasize the importance of keeping ladders clean. Store ladders properly and in areas away from pedestrian traffic. Keep safety feet and other parts in good condition. Require users to inspect each ladder before beginning work. Repair ladders with defects or immediately remove them from service. Place a danger tag on each defective ladder. OSHA General Industry Standards found in 29 CFR 1910.22, 23, and 28 address "nonconstruction" scaffolding and platforms. A trained supervisor must inspect all platforms or scaffolds before use. All elevated platforms must contain standard guardrails securely fastened to a stationary object.

Control of Hazardous Energy (29 CFR 1910.147)

Develop and implement appropriate procedures that render inoperative any source containing hazardous energy. Sources can include electrical systems, pumps, pipelines, valves, and tanks. Employers must develop written procedures for lockout/tagout procedures. They must also conduct employee training, ensure accountability of engaged employees, and develop appropriate administrative control. The employer must maintain a list of authorized lockout or tagout personnel and specify any restrictions. Develop procedures for each machine that contains more than a single energy source. Train authorized personnel to ensure knowledge of all procedures. Evaluate procedures and retrain annually. Mandatory OSHA requirements include the following:

- Written procedures for lockout and tagout
- Training of employees
- Accountability of engaged employees
- Establishment of administrative controls

Educate all employees on the importance of respecting lockout and tagout devices. Only qualified personnel can remove control devices. Training must ensure that employees understand the purposes, functions, and restrictions of the energy control procedures. Employers must also conduct training specific to the needs of authorized, affected, and other employees. Authorized employees can implement energy control procedures or perform service or maintenance activities. Authorized employees need training to address hazardous energy source recognition, types and magnitude of the hazardous energy sources in the workplace, and energy control procedures, including the methods to isolate and control energy sources (Table 4.7).

OSHA defines "affected employees" as those who do not service or maintain machinery or perform lockout/tagout activities. However, these employees work in areas near hazardous energy sources. Affected employees must receive training in the purpose and use of energy control procedures. They must understand the importance of not tampering with any lockout or tagout devices. Prohibit these employees from starting or using equipment locked out or tagged. Employees working in areas with energy control procedures in place must receive instruction regarding the energy control procedures and the prohibition against removing a lockout or tagout device.

Table 4.7   Lockout Basic Steps

- Prepare for shutdown.
- Shut off the machine and/or energy supplies.
- Disconnect or isolate the machine from the energy sources.
- Apply the lockout or tagout devices to the energy-isolating devices.
- Release, restrain, or render safe potential hazardous stored/residual energy.
- Ensure accumulated energy does not reaccumulate to hazardous levels.
- Verify the isolation and degeneration of the machine.

Employers must provide retraining for all "authorized and affected employees" when a change occurs in job assignments, machinery or processes, or energy control procedures. Conduct retraining when a periodic inspection or observation reveals inadequate procedures or an employee shows a lack of knowledge. Employers must review procedures at least annually to determine adequacy to protect. Employers must correct deviations and inadequacies identified in the energy control procedures or in its application. Employees must also receive retraining when a change occurs in their job assignment.

Permit Confined Spaces (29 CFR 1910.146)

Evaluate all confined spaces. Consider any enclosed area as hazardous until properly tested. Conduct testing for oxygen deficiency and combustible-gas indicators prior to entry. Never enter atmospheres containing 19.5% or less of oxygen by volume without the use of an air-supplied respirator. OSHA revised the standard in 1998 to provide for more employee participation in permit-confined space procedures. It authorized representatives with the opportunity to observe any testing or monitoring of permit spaces. OSHA defines a confined space as "any space" with limited or restricted means of entry or exit but large enough for employees to enter and perform work. In addition, the design must prohibit continuous occupancy. OSHA also specifies requirements for nonpermit confined spaces. These spaces cannot contain hazardous atmospheres and pose no risk of causing death or serious harm. If a space receives designation as permit-required, the employer must inform all exposed employees of the dangers by posting signs or using some another equally effective means. Employers must establish and implement procedures to prevent unauthorized entry. Employers must also eliminate or control hazards necessary for safe entry by specifying acceptable entry conditions and isolating the space. Employers must purge, make inert, flush, or continuously ventilate the permit space to eliminate or control atmospheric hazards (Table 4.8).

Employers must provide, maintain, and require use of PPE as required for safe entry. OSHA requires training to ensure that employees involved in confined space work can perform job functions safely. The training must cover specific items for the authorized entrant, the attendant, and the entry supervisor. OSHA now recognizes the right for employees or their authorized representative to participate in the development and implementation of permit space procedures. They can observe any periodic testing and review the certification of preentry safety measures. OSHA permits employees to review employer documentation that all hazards in a permit space have been eliminated. They now have the right to review the results of any testing conducted and the completed permit.

Table 4.8 Permit-Required Confined Space Characteristics

- Contains or could contain a hazardous atmosphere
- Contains material that has the potential for engulfing the entrant
- Inwardly converging walls
- Any other recognized safety or health hazards

System Safety and Engineering

Two early pioneers in system thinking, Daniel Katz and Robert Kahn, viewed most organizations as open social systems. These open systems consisted of specialized and interdependent subsystems. These subsystems possessed processes of communication, feedback mechanisms, and management intervention that linked them together. Katz and Kahn held that the closed system approach failed to take into account how organizations reciprocate and depend on external environments. For example, environmental forces such as customers and competitors exert considerable influence on corporations. This influence highlights the essential relationship between an organization and its environment. That relationship promotes the importance of maintaining external inputs to achieve organizational stability. We can describe a system using three brief definitions: (1) a set of interrelated parts that function as a whole to achieve a common purpose, (2) a piece of software that operates to manage a related collection of tasks, or (3) a design for an organization that perceives sets of processes as a related collection of tasks.

Systems operate as either open or closed entities or processes. Systems can take various forms or shapes and express themselves as mechanical, biological, or social entities. Open systems can interact with other inside subsystems or the outside environment. Closed systems exert little interaction with other systems or the outside environment. Open systems theory originated in the natural science fields. It subsequently spread to fields as diverse as information technology, engineering, and organizational management. Open systems view an organization as an entity that takes inputs from the environment, transforms those inputs, and releases outputs. This results in reciprocal effects on the organization itself along with the environments in which the organization operates. An organization can become part of the environment in which it operates. The majority of systems operate as open entities. These systems require interaction with the environment for the source of inputs and the destination of outputs (Tables 4.9 and 4.10).

Safety engineering addresses fundamental principles and rules used to identify, evaluate, and control hazards within "man and machine" interface systems. Consider safety engineering as the physical and mathematical components of loss-control efforts. We must view hazard control management as an organizational function that addresses the leadership, behavioral, and administrative aspects of preventing loss. Hazard control management must provide the structure and insight for applying safety engineering and human factors principles to accident prevention. Safety engineers analyze the early design of a system to discover potential faults and flaws.

Safe design attempts to achieve an acceptable mishap risk through a systematic application of guidance obtained from standards, specifications, regulations, handbooks, checklists, and other sources. Safe-design needs derive from the selected parameters and associated acceptance criteria. The life cycle of systems includes design, research, development, evaluation, production, inventory, operational support, and disposal. Probabilistic fault tolerance adds redundancy to equipment and systems.

Table 4.9 Design-Related Weaknesses

- Failure to design adequately
- Missing or inadequate policies and rules
- Training and education objective not developed
- Poorly written plans
- Inadequate processes
- Lack of appropriate procedures

Table 4.10 Operational-Related Failures

- Not implementing or carrying out required functions or processes
- Not adhering to established policies, procedures, and directives
- Not developing and presenting appropriate education and training sessions
- Not ensuring adequate supervision, or failing to provide required oversight
- Not conducting comprehensive accident or root-cause analysis after a mishap
- Not evaluating plans, procedures, and processes to determine weaknesses

The expression *safety critical* refers to any condition, event, operation, process, or item whose proper recognition, control, performance, or tolerance is essential to safe system operation. The underlying concept of system safety considers the whole as more than the sum of its parts. System processes focus on eliminating, controlling, and managing hazards throughout the life cycle of the system. The following steps contribute to safe operational designs: (1) planning, (2) hazard identification, (3) hazard analysis, (4) risk assessment, and (5) making proper decisions through use of risk management to implement suitable controls. System-safety techniques can help control accidents and loss by focusing on discovering the underlying causal factors. System safety must accept an optimized level of risk constrained by cost, time, and operational effectiveness/performance. *Normative safety* describes products that meet applicable design specifications or standards. We can define a *mishap* (another term for accident) as an unplanned event or series of events resulting in death, injury, illness, damage to or loss of equipment or property, or damage to the environment. Mishap risk relates to the potential severity and probability of occurrence. Mishap probability relates to an arbitrary categorization that provides a qualitative measure of the most reasonable likelihood of occurrence.

This undesired event can result from personnel error, environmental conditions, design inadequacies, and procedural deficiencies. It can also relate to system, subsystem, or component failure or malfunction. System-safety methods require acceptance of some level of mishap risk, determine mishap probability, establish severity threshold, and create appropriate documentation procedures (Table 4.11).

System reliability refers to the probability that a system or process will consistently perform as designed. We can view reliability as opposite of a system's rate of error or failure. Many reliable systems will not work effectively unless accepted in an organizational culture that values teamwork, communication, accountability, and learning from mistakes. Many improvement processes focus too much on operational effort expended instead of focusing on process or system-design issues.

Table 4.11   Key System Safety Elements

• Standardize, simplify, and automate as much as possible.
• Minimize fatigue, stress, and boredom.
• Reduce reliance on human memory, but promote human vigilance.
• Encourage teamwork, and improve reporting accuracy/timeliness.
• Enhance information transfer within the organization.
• Design equipment to reduce failures.
• Consider technology and the human interface when designing processes.
• Study history to ensure that patient safety continues to improve over time.
• Statistics can help measure the impact of interventions or innovations.
• Continuous improvement processes shift focus from an individual to a team.

Table 4.12   System Failure and Error Considerations

• Complex systems can "break or fail" and contribute to harm.
• Systems and processes may contain a latent defect that leads to harmful results.
• Humans tend to develop or modify behaviors to compensate for chronic system flaws.
• Errors can occur far from the actual operational location of the system or process.
• Some organizational cultures can promote too much individual accountability.
• Those attempting to fix system errors may not see how their actions impact operations.
• Personnel can spend valuable time trying to determine factors that contribute to harm.
• Personnel must understand that any part of a system can impact operational integrity.

Too much focus on individual outcomes can exaggerate overall reliability. Autonomy permits systems to have wide margins of performance.

When designing system processes, focus on meeting specific goals and objectives.

Reliability-centered maintenance (RCM) focuses on analysis of potential failures. RCM methodology, such as used with aircraft, involves failure modes related to serviceable and/or replaceable components. This process helps predict an impending failure of a component. Complex systems can experience a large number of maintenance requirements, as identified by failure analysis.

The RCM analysis, when used to identify maintenance actions, can reduce the probability of failure with the least amount of cost. This includes using monitoring equipment for predicting failure. RCM relies on up-to-date operating performance data compiled from a computerized maintenance system source. The data collected proves valuable when used in a failure mode, effects, and criticality analysis (FMECA) process to rank and identify the failure modes of concern (Table 4.12).

System-Related Discovery Tools

Hazard and operability (HAZOP) studies can help systematically investigate each element of an operational system or process. HAZOP studies can help discover ways in which important parameters can deviate from intended designs or configurations. Deviations can create hazards and other operational problems. An expert team can address HAZOP problems using diagrams to study the effects of such potential hazards. The team must seek to identify and then select the parameters needing study. The team must assess the impact of design deviations. The team agrees on a list of descriptive key words such as *more of* or *part of* to describe each noted deviation. The team evaluates the system as designed. However, the team also notes all deviations that could cause failure. The team then seeks to identify effectiveness of any existing controls or protective safeguards. Finally, the team must evaluate causal factors, determine the consequences of failure, and determine the corrective actions needed to control or eliminate the hazards.

FMECA uses a diagram of a system or process to determine potential failures and resulting severity consequences. System safety personnel must consider failure outcomes at each block or point of the diagramed system. The analysis permits assessment of potential failures using frequency-of-occurrence empirical data. The process would evaluate and document by failure mode the potential impact of each functional or hardware failure. Such failures could impact system success, safety, maintainability, and performance. If possible, consider analyzing multiple concurrent failures during an analysis process. Consider each component's effect on operational safety and impact of failure. An effective analysis must consider both the frequency and severity of possible component failure. Analyze each component to determine potential mode of failure, effects of failure, and failure-detection methods. The final aspect of the FMECA process consists of analyzing component data to develop hazard controls that lower risks and consequences of failure. The last step in the process involves the analysis of the data for each component or multiple component failure. Finally, develop recommendations that address appropriate risk-management action or intervention.

FMECA helps identify parts, processes, or systems that are most likely to fail. Using FMECA during the design phase can reduce overall costs by identifying single-point failures and other concerns prior to system development. FMECA works well as a troubleshooting tool to identify corrective actions for a potential failure. FMECA addresses the areas of design, operating parameters, and maintenance of the system. It will address system safety challenges, risks, and problems.

FMECA processes can discover and document issues in the following areas: (1) design and development, (2) manufacturing, (3) complex processes, (4) safety and hazard control, and (5) environment-related concerns. Human harm remains the top consequence of failure (Table 4.13).

Fault-tree analysis (FTA) focuses on the identification of multiple point failures by using a deductive top-down method to analyze effects of initiating faults and events occurring in complex systems. FTA works very well, showing how complex systems can overcome single or

Table 4.13 Traditional Severity Classifications

- Category I-Catastrophic—A failure that may cause injury or death
- Category II-Critical—A failure which may cause severe injury, major property damage, or major system damage that will result in major downtime or production loss
- Category III-Marginal—A failure which may cause minor injury, minor property damage, or minor system damage which will result in delay or loss of system availability or degradation
- Category IV-Minor—A failure not serious enough to cause injury, property damage or system damage, but will result in unscheduled maintenance or repair

multiple initiating faults that could result in failure. FTA does not work well in discovering all possible initiating faults. More than one condition must occur for a particular failure to happen. The probability of failure for various components uses Boolean algebraic symbols. Determine and place on the fault tree each situation or hazard that could impact the system. Do this by using a series of logic expressions. Fault trees using actual numbers permit the calculation of failure probabilities. Event trees can start from an undesired initiator, such as the loss of critical power supply or component failure. Event trees follow a fault through other system components. This allows for determination of final consequences. When considering a new event, add a new node on the tree. This permits a splitting of probabilities by taking either branch. FTA considers external events, whereas FMECA does not. FTA starts by examining an undesirable outcome and then "traces back" through the diagram to identify all possible events or combinations of events that would need to occur to produce *that* specific outcome.

FTA requires the use of logic symbols to trace the sequences of events that could result in an incident. The resulting diagram looks like a tree with many branches, with each branch listing sequential events or failures for different independent paths to the top event. Analysts then can better assign failure rate data to each event and calculate the probability of an undesired event occurrence.

Risk management in any setting can be described as the probability that a hazard will cause injury or damage. In some organizations, risk management operates separately from the hazard control function. For example, hospitals consider risk management to be a separate function from environmental safety efforts. Some other types of organizations may consider risk management as an integral element of hazard control function. Risk management from an insurance and loss-control perspective can quickly become a reactive managerial element. Risk management views *all* losses, not just human injury, to the organization. Risk assessment relates to the process by which risk-analysis results drives decision making. Risk-control efforts address hazardous events by implementing interventions to reduce severity. Risk management includes not only control efforts but finance as well. Risk control considers all aspects of system safety, hazard control management, and safety engineering. Risk finance considers insurance, risk pooling, and self-insurance (Table 4.14).

Instructional systems design (ISD) can maximize effectiveness, efficiency, and appeal of instruction. The process must determine the current knowledge or competence of the learner. The process must also know the level of competence the learner must achieve during the instructional process. ISD can help define learning objectives, and it uses proven instructional methods to achieve results. The basic model works in most training environments. The military used some basic pre-ISD concepts during World War II. Leaders broke down important tasks and subtasks. They treated each one as a separate learning goal. Training rewarded correct performance and

Table 4.14   Risk Management Process

• Step 1: Identify hazards using the effective tools and techniques.
• Step 2: Assess risks to life, property, and the environment.
• Step 3: Develop controls and make decisions on what risks are acceptable.
• Step 4: Implement controls and track corrective actions through completion.
• Step 5: Periodically evaluate the effectiveness of the risk management process.

Table 4.15   System Functions

- Management—directing/controlling instructional system development and operations
- Support—maintaining or servicing all parts of the system
- Administration—day-to-day processing and record keeping
- Delivery—bringing and delivering instruction to students
- Evaluation—obtaining feedback and information through use of formative, summative, and operational evaluations to assess system and student performance

used remediation processes to address incorrect performance. ISD concepts developed from contributions from many disciplines, including system engineering, behavioral science, cognitive psychology, and instructional methodology (Table 4.15).

Industrial Hygiene

Industrial hygienists (IHs) analyze, identify, and measure workplace hazards or stressors that can cause sickness, impaired health, or significant discomfort in workers through chemical, physical, ergonomic, or biological exposures. IHs use environmental monitoring and analytical methods to detect the extent of worker exposure and employ engineering, work-practice controls, and other methods to control potential health hazards.

Typical roles of the IH include investigating and examining the workplace for hazards and potential dangers. An IH also makes recommendations on improving the safety of workers and the surrounding community. IH professionals conduct scientific research to provide data on possible harmful conditions in the workplace. They educate and train the community on a variety of job-related risks. IHs deal with the health and safety challenges facing people everywhere, including (1) indoor air quality, (2) lead exposure, (3) emergency response planning and community right-to-know activities, (4) tuberculosis and silicosis, (5) providing recommendations to reduce exposure to potentially hazardous agents, (6) cumulative trauma disorders and repetitive stress injuries, and (7) evaluation of hazards associated with radiation and electromagnetic fields. IHs also work in the areas of controlling reproductive health hazards and reducing exposures to hazardous substances in the workplace. IHs work to detect and control occupational noise, radiation, and illumination.

Ergonomics and Human Factors

The word *ergonomics* comes from the Greek words *ergo* or work and *nomos* or law. It can also be referred to as the science or art of fitting the job to a worker. A mismatch between the physical requirements of a task and the physical capacity of the worker can result in a musculoskeletal disorder. The International Ergonomics Association defined ergonomics in 2000 as follows: "a scientific discipline concerned with the understanding of interactions among humans and other elements of a system and the profession that applies theory, principles, data, and methods to design in order to optimize human well-being and overall system performance." Ergonomics and human factors are often used interchangeably. Ergonomic hazards refer to workplace conditions that pose the risk of injury to the musculoskeletal system of the worker. Ergonomic hazards include repetitive and forceful movements, vibration, temperature extremes, and awkward postures that arise from improper work methods and improperly designed workstations, tools, and equipment. Ergonomics addresses issues related to the "fit" between people and their technological tools and environments.

Ergonomics draws on many disciplines in its study of humans and their environments, including anthropometry, biomechanics, mechanical engineering, industrial engineering, industrial design, kinesiology, physiology, and psychology. Many organizations develop and implement an ergonomics policy with written goals, objectives, and accountability policies. Leaders should encourage worker involvement in the ergonomics improvement efforts. NIOSH recommends reducing or eliminating potentially hazardous conditions using engineering controls or implementing work practices and improved management policies. To meet ergonomics challenges, equipment should comply with ergonomics principles.

Effective training covers the problems found in each employee's job. Training programs can go a long way toward increasing safety awareness among both managers and employees. Training and education can ensure employees are sufficiently informed about workplace hazards. Soliciting suggestions from workers about ergonomic hazards can help improve work practices. A good training program properly instructs employees to use equipment, tools, and machine controls.

Reactive ergonomics takes only corrective actions when required to do so by injury or complaint. Proactive ergonomics seeks to identify all areas needing improvement. Attempt to solve problems by changing equipment design, modifying job tasks, and improving environmental designs. Health-care providers need to be familiar with worker, jobs, and tasks and participate in matching jobs and work environments to worker needs. Use information obtained from job-hazard analyses, job descriptions, photographs, and videotapes to identify ergonomic hazards. According to the International Ergonomics Association, physical ergonomics addresses human anatomical, anthropometric, and physiological issues that relate to physical activity. Cognitive ergonomics addresses the concern with mental processes such as perception, memory, reasoning, and motor response. Macroergonomics emphasizes a broad system view of design, considering organizational environments, culture, history, and work goals. It deals with the physical design of tools and the environment. It is the study of the society and technology interface and considers human, technological, and environmental variables and their interactions (Table 4.16).

Human factors, science, and ergonomics are many times used interchangeably. Human-factors science incorporates contributions from psychology, engineering, industrial design, statistics, operations research, and anthropometry. Human factors as a science covers the science of understanding the properties of human capabilities. The application of this understanding to the design,

Table 4.16 Examples of Ergonomic Risk Factors

- Jobs requiring identical motions every 3 to 5 seconds for more than 2 hours
- Work postures such as kneeling, twisting, or squatting for more than 2 hours
- Use of vibration or impact tools or equipment for more than a total of 2 hours
- Lifting, lowering, or carrying more than 25 pounds more than once during the work shift
- Piece-rate or machine-paced work for more than 4 hours at a time
- Workers' complaints of physical aches and pains related to their work assignments

Table 4.17 Factors Impairing Human Performance

- Limited short-term memory
- Running late or being in a hurry
- Inability to multitask
- Interruption of the job or task
- Stress or the lack of sleep
- Fatigue or effects of shift work
- Environmental factors
- Personal or home distractions
- Drug and substance abuse

development, and deployment of systems and services relates to human-factors engineering. Human factors can include sets of human-specific physical, cognitive, or social properties. These human-factor sets can interact in a critical or dangerous manner with technological systems, the human natural environment, or human organizations. Human-factors engineering applies knowledge about human capabilities and limitations to the design of products, processes, systems, and work environments. It also relates to the design of all systems having any type of human interface. Its application to system design improves ease of use and performance while reducing errors, operator stress, training, user fatigue, and product liability. It is the only discipline that relates humans to technology. Human-factors engineering focuses on how people interact with tasks, machines or computers, and the environment, with the consideration that humans have limitations and capabilities (Table 4.17).

Robotic Safety

International Standards Organization (ISO) defines industrial robots as automatically controlled, reprogrammable, or multipurpose manipulators with programming in three or more axes. Typical applications of robots include welding, painting, assembly, pick and place, packaging, palletizing, product inspection, and testing. An industrial robot system includes not only industrial robots but also any devices and sensors required for the robot performance. This definition includes sequencing, monitoring, or communication-related interfaces. Most robots use a teach-and-repeat technique. A trained operator typically uses a portable control device known as a teach pendant to instruct the robot how to accomplish its task. Some robots received programming to conduct specific repetitive actions accurately without variation. These actions, determined by programs, specify the direction, acceleration, velocity, deceleration, and distance of a series of coordinated motions. The working envelope consists of the region of space a robot can reach. Kinematics refers to the actual arrangement of rigid members and joints in the robot that can determine possible motions.

The payload relates to how much weight a robot can lift, and speed relates to how fast a robot can position the end of its arm. The acceleration of a robot refers to how quickly an axis can accelerate, and accuracy relates to how closely a robot can reach a commanded position. We can define repeatability as how well the robot will return to a programmed position.

Studies show that many robot accidents do not occur under normal operating conditions. They occur during programming, maintenance, testing, setup, or adjustment. During these operations, an operator, a programmer, or a maintenance worker may temporarily be within the working envelope where unintended operations could result in injuries. Typical accidents can include events such a robotic arm malfunctioning during a programming sequence and striking the operator, or someone tripping a power switch while maintenance was under way, resulting in a hand injury. The proper selection of effective robotic safeguarding systems must be determined by using hazard analysis of system use, programming, and maintenance operations.

An effective safeguarding system will protect operators, engineers, programmers, and maintenance personnel. Recommend the use of a combination of safeguarding methods to include redundancy and backup systems. The safeguarding devices employed should not create a hazard or curtail necessary vision. Unpredicted movements, component malfunctions, or program changes related to the robot's arm or peripheral equipment could result in contact accidents. A worker's limb or other body part can be trapped between a robot's arm and other peripheral equipment. Environmental incidents can result from an arc flash, metal spatter, dust, or electromagnetic interference. Pneumatic, hydraulic, or electrical power sources that have malfunctioning control or transmission elements in the robot power system can disrupt electrical signals to the control and/or power-supply lines. Fire risks are increased by electrical overloads or by use of flammable hydraulic oils. Electrical shock and release of stored energy from accumulating devices also can be hazardous to personnel.

All robotic machinery should comply with requirements of Section 4 of the ANSI/RIA R15.06-1992, Standard for Manufacturing, Remanufacture, and Rebuild of Robots. A robot or robot system should be installed by the users in accordance with the manufacturer's recommendations and in conformance to acceptable industry standards. Temporary safeguarding devices and practices should be used to minimize the hazards associated with the installation of new equipment. A key consensus standard that addresses robotic safety is the Industrial Robots and Robot Systems (Safety Requirements ANSI/RIA R15.06-1999). ANSI defines singularity as a condition caused by the collinear alignment of two or more robot axes resulting in unpredictable robot motion and velocities. Another safety development includes the downsizing of industrial arms for light industrial uses.

In addition, OSHA's Lockout/Tagout standards (29 CFR 1910.147 and 1910.333) must be followed for servicing and maintenance. At each stage of development of the robot and robot system, a risk assessment should be performed. There are different system and personnel safeguarding requirements at each stage. The appropriate level of safeguarding determined by the risk assessment should be applied. In addition, the risk assessments for each stage of development should be documented for future reference. Special consideration must be given to the teacher or person who is programming the robot. The system operator should be protected from all hazards during operations performed by the robot. Safeguarding maintenance and repair personnel is very difficult, because their job functions are so varied. Troubleshooting faults or problems with the robot, controller, tooling, or other associated equipment is just part of their job. Personnel who program, operate, maintain, or repair robots or robot systems should receive adequate safety training, and they should be able to show their competence to perform Safety and Health Administration, OSHA 29 CFR 1910.333, Selection and Use of Work Practices, and OSHA 29 CFR Part 1910.147, The Control of Hazardous Energy (Lockout/Tagout) (Table 4.18).

Table 4.18   Key Robotic Safety Consensus Standards

ANSI R15.06-1999, Industrial Robots and Robot Systems, Safety Requirements—provides requirements for industrial robot manufacture, remanufacture and rebuild, and robot system integration/installation and methods of safeguarding to enhance the safety of personnel associated with the use of robots and robot systems.
TR R15.106-2006, Technical Report on Teaching Multiple Robots, Robotics Industries Association (RIA)—provides additional safety information relative to teaching (programming) multiple industrial robots in a common safeguarded space in an industrial setting, and supplements the ANSI/RIA R15.06-1999 robot safety standard.
B11.TR3-2000, Risk Assessment and Risk Reduction, A Guide to Estimate, Evaluate and Reduce Risks Associated with Machine Tools—provides a means to identify hazards associated with a particular machine or system when used as intended, and provides a procedure to estimate, evaluate, and reduce the risks of harm to individuals associated with these hazards under the various conditions of use of that machine or system.
International Organization for Standardization (ISO) TC 184, Industrial Automation Systems and Integration ISO 10218-1:2006, Robots for industrial environments, Safety requirements, Part 1: Robot, Robotics Industries Association (RIA)—specifies requirements and guidelines for the inherent safe design, protective measures, and information for use of industrial robots; describes basic hazards associated with robots; and provides requirements to eliminate or adequately reduce the risks associated with these hazards.
Canadian Standards Association (CSA) Z434-03, Industrial Robots and Robot Systems—applies to the manufacture, remanufacture, rebuild, installation, safeguarding, maintenance and repair, testing and start-up, and personnel training requirements for industrial robots and robot systems.
American Welding Society (AWS) D16.1M/D16.1, Specification For Robotic Arc Welding Safety—identifies hazards involved in maintaining, operating, integrating, and setting up arc-welding robot systems.

Environmental Management

Environmental management is more than EPA compliance or managing hazardous materials. The three main issues that affect environmental managers are those involving politics, projects, and resources. The need for environmental management can be viewed from a variety of perspectives. Environmental management is therefore not the conservation of the environment solely for the environment's sake. It is conservation of the environment for humankind's sake. It focuses on the solution of the practical problems that humans encounter in cohabitation with nature, exploitation of resources, and production of waste. The agents of environmental management can include foresters, conservationists, policy makers, engineers, and resource planners.

There is environmental impact analysis, which is codified in the U.S. National Environmental Policy Act. Through the environmental impact statement, the analysis prescribes the investigation and remedial measures that must be taken to mitigate the adverse effects of new development. It is intended to act in favor of both prudent conservation and participatory democracy.

Audits use techniques such as life-cycle analysis and environmental-burden analysis to assess the impact of, for example, manufacturing processes that consume resources and create waste. Environmental management involves the management of all components, both living (biotic) and nonliving (abiotic), of the biophysical environment. This is due to the interconnected network of relationships among all living species and their habitats. The management of the environment also involves the relationships of the human environment, including social, cultural, and economic issues.

The ISO 14001 Standard remains the most widely used standard for environmental risk management. The standard is closely aligned to the European Eco-Management and Audit Scheme (EMAS). As a common auditing standard, ISO 19011 Standard explains how to combine this with quality management. Other environmental management systems (EMSs) tend to be based on the ISO 14001 standard.

Product Safety

The basic elements of product safety management have been generally presented in various authoritative texts, beginning with general safety engineering and, in recent years, with works dedicated to product safety management. The American National Standards Institute's "Guidelines for Organizing a Product Safety Program" states that manufacturers have a responsibility to produce products that satisfy the safety expectations of society. The public does have special expectations of engineers, designers, and others trained to improve safety for the rest of society. Professional responsibility is based on the belief that the power conferred by expertise entails a fiduciary relationship to society, as presented by William W. Lowrance in his text *Of Acceptable Risk*, published in 1976. Product safety involves the application of the principles of safety management and engineering to the design and marketing of products.

Basic elements of product safety programming are designed to identify and evaluate potential product hazards for systematic control, using the techniques of safety management and safety engineering (Table 4.19).

Table 4.19 Basic Product Safety Principles

- Integrate safety into every facet of product design.
- Educate consumers about key issues that relate to product safety.
- Develop product-safety testing protocols for all foreseeable hazards.
- Keep current and implement, where feasible, the latest developments in product safety.
- Work proactively and cooperatively with the Consumers Products Safety Commission.
- Conduct thorough and comprehensive investigation of all product-safety incidents.
- Promptly initiate product recall procedures on discovery of a product-safety issue.
- Track and address product's safety performance, and investigate product incidents.
- Integrate systems safety principles in the design and development of products.

An effective product safety function must formally declare to all personnel that product safety is important. Leaders must assign responsibility to specific individuals to assure product safety during the product design, manufacturing, and marketing process. Establish specific activities to identify and evaluate potential product hazards based on reasonably foreseeable conditions of product use. Utilize reasonable well-established and available safety standards or guidelines to design safely, safeguard remaining hazards, and provide adequate product warnings or instructions to address hazards that remain. Develop a concise product safety policy to include the assignment of individual responsibility for the function.

Implement a well-written plan outlining the specific steps, procedures, and techniques to achieve safe design. Obtain authoritative literature, texts, and relevant standards relating to product safety and marketing. Conduct and document actions taken to identify reasonably anticipated or potential product or system hazards. Analyze and evaluate identified hazards and determine associated risk factors. Document use of system- and safety-management principles and engineering to reasonably eliminate or minimize unacceptable product hazards through design, safeguarding, or warning. Assign responsibility for product control and compliance to specific individuals, and ensure they understand their role in the organization. Ensure that the assigned individuals have the necessary training, knowledge, experience, skills, and competence to perform product compliance and control responsibilities. Maintain appropriate control of activities by having clearly documented policies, specifications, and procedures to ensure product safety, and maintain records to show how compliance was achieved. Establish a process to analyze and evaluate risks in the product life-cycle, and develop an approach to control those risks appropriately.

Develop and maintain a system for communication and information that allows the sharing of relevant information on safety and compliance within the organization and with third parties, including federal, state, and local authorities. Establish a formal, documented quality assurance program designed to control, monitor, and improve operations continually to ensure the safety of products and compliance with all applicable U.S. requirements.

The International Board for Certification of Safety Managers (IBFCSM) offers qualified candidates the opportunity to earn the Certified Product Safety Manager (CPSM) credential. Qualified candidates need at least 8 years of education and/or experience combined to sit for the examination. Refer to the *CPSM Handbook* found on the board's website at www.ibfcsm.org for additional information and the examination outline.

The Consumer Product Safety Commission (CPSC), an independent federal regulatory agency, was created in 1972 by Congress in the Consumer Product Safety Act. In that law, Congress directed the commission to "protect the public against unreasonable risks of injuries and deaths associated with consumer products." CPSC exercises jurisdiction over about 15,000 types of consumer products. The CPSC's work to ensure the safety of consumer products such as toys, cribs, power tools, cigarette lighters, and household chemicals have contributed significantly to the 30% decline in the rate of deaths and injuries associated with consumer products over the past 30 years.

The Consumer Product Safety Improvement Act of 2008 (Public Law 110-314) increased the budget of the CPSC, imposed new testing and documentation requirements, and set new acceptable levels of several substances including lead. It imposes new requirements on manufacturers of apparel, shoes, personal care products, accessories and jewelry, home furnishings, bedding, toys, electronics and video games, books, school supplies, educational materials, and science kits. The act also increased fines and specifies jail time for some serious violations (Table 4.20).

Table 4.20   Key Product Safety Definitions

Hazard: The ability of a substance to harm or produce adverse effects. In the context of our businesses, hazards may be associated with raw materials and manufacturing processes, product packaging, or use and foreseeable misuse (or any misuse) of our products. Hazards may be health hazards, physical or chemical hazards, or environmental hazards.
Product labels and safety data sheets: These documents, in general, identify the known hazards associated with a particular substance. Exposure refers to the conditions of the interaction between a person(s) and a substance/product. Important aspects of exposure are the duration of exposure, the amount of exposure, and the route of the exposure.
Risk: This is the probability that an adverse effect will occur under specific conditions of exposure, such as normal use or reasonably foreseeable misuse. It will be readily seen that risk is a function of both hazard and exposure. Most importantly, there is no risk without exposure, no matter how hazardous a substance may be.
Product risk assessment: The process by which the risk of harm associated with a specific substance and specific exposure conditions is estimated. Risk assessments, where possible, are comprised of four components: (1) hazard identification, (2) dose response determinations, (3) exposure assessments, and (4) risk characterization.
Product risk management: The process that deals with unacceptable levels of risk and seeks to determine what controls are necessary.
Product safety: The probability that harm is unlikely to occur under specific conditions of use, including normal use or reasonably foreseeable misuse. A safe product is one that either presents no adverse health effects or only minimal risk of adverse health effects. It must be compatible with the product's chemistry, required efficacy, and use when used as directed or under reasonably foreseeable conditions of misuse.

Hazard Analysis Control Critical Point

Foodborne disease or illness usually occurs within a few hours to a few weeks among individuals who have eaten the same food. The detection of contaminating agents, using taste, odor, texture, or appearance, can prove difficult. Hazard Analysis Control Critical Point (HACCP) is one of the most widely recognized and used tools in the food industry to control the risk of food contamination. HACCP is also a management system in which food safety is addressed through the analysis and control of biological, chemical, and physical hazards from raw material production, procurement, and handling to manufacturing, distribution, and consumption.

HACCP principles are endorsed by the U.S. Department of Agriculture's Food and Safety Inspection Service (FSIS) and the Food and Drug Administration (FDA). HACCP helps to prevent, as close to 100% as possible, any harmful contamination of the food supply. Potential hazards could be biological, such as a microbe, or chemical, such as a toxin. They can be physical, such as ground glass or metal fragments. The objective is to identify critical control points during a food's production from its raw state through processing, shipping, and consumption by the consumer. Examples include cooking, cooling, packaging, and metal detection during the process. Establish preventive measures with critical limits for each control point. For a cooked food, for example, this might include setting the minimum cooking temperature and time required to ensure the elimination of any harmful microbes. Establish procedures to monitor the critical

Table 4.21   HACCP Elements

> - Identify and analyze workplace hazards—identify potential food-related hazards, and suggest appropriate control measures.
> - Identify critical control points—a critical control point is defined as a point, procedure, or step at which a food-safety hazard could be eliminated, prevented, or reduced.
> - Establish critical limits—critical limits refer to maximum and/or minimum values that prevent, eliminate, or reduce a biological, chemical, or physical hazard.
> - Monitor the critical control points—ensure monitoring remains continuous to assure accurate readings.
> - Define corrective actions—determine actions to correct the cause of deviation, and establish whether a food product should be discarded.
> - Establish verification procedures—verification activities determine the effectiveness of the plan and ensure correct system operation.
> - Establish recordkeeping and documentation procedures—documentation should be kept for the entire food process and should include a summary of the hazard analysis.

control points. Such procedures might include determining how and by whom cooking time and temperature should be monitored. Establish corrective actions to be taken when monitoring shows that a critical limit has not been met—for example, reprocessing or disposing of food if the minimum cooking temperature is not met. Establish procedures to verify that the system is working properly. For example, use time- and temperature-recording devices to verify that a cooking unit is working properly.

Establish effective record keeping documenting the HACCP system. This would include records of hazards and their control methods, the monitoring of safety requirements, and action taken to correct potential problems. Each of these principles must be backed by sound scientific knowledge: for example, published microbiological studies on time and temperature factors for controlling foodborne pathogens. New challenges to the US food supply have prompted FDA to consider adopting a HACCP-based food-safety system on a wider basis. One of the most important challenges is the increasing number of new food pathogens. For example, between 1973 and 1988, bacteria not previously recognized as important causes of foodborne illness—such as *Escherichia coli* O157:H7 and *Salmonella*—became more widespread. There also is increasing public health concern about chemical contamination of food. A good example would be the effects of lead in food on the human nervous system. Another important factor is that the size of the food industry and the diversity of products and processes have grown tremendously in the amount of domestic food manufactured and the number and kinds of foods imported (Table 4.21).

Laboratory Safety

Laboratories operate in a variety of settings, such as hospitals, dental operations, research facilities, secondary schools and colleges, and governmental facilities. Many industrial companies operate laboratories to address issues such as quality assessment or product development. Laboratories handling biohazards must take additional actions to prevent exposure of employees to blood-borne pathogens. Laboratories can expose workers to a variety of other hazards, including chemicals, toxins, and flammable materials. The OSHA Laboratory Standard, 29 CFR 1910.1450, is performance oriented and allows employers the flexibility to implement specific safe work practices. The standard covers laboratory workers in industrial, clinical, and school settings.

The standard covers all chemicals that meet the definition of a health hazard as defined in the OSHA Hazard Communication Standard. The written chemical hygiene plan provides the basis for meeting the requirements of the standard. The plan establishes appropriate work practices, standard operating procedures, control methods, and measures for appropriate maintenance and use of protective equipment. The plan must also detail procedures regarding medical examinations and special precautions for work with particularly hazardous substances. The plan must be reviewed annually and updated as required. The written program must be available to employees and their designated representatives. Employers must appoint a chemical hygiene officer, and, as appropriate, establish a chemical hygiene committee. Establish a training and information program for employees exposed to hazardous chemicals in the workplace. The worker program should be initiated at the time of initial assignment and prior to assignments involving new exposure situations. This provision incorporates the training and information requirements of the hazard communication standard. Training topics include location of the facility hygiene plan and requirements of the OSHA Laboratory Standard.

The laboratory standard does not mandate medical surveillance for all laboratory workers. The employer must provide workers an opportunity for medical attention. This includes follow-up examinations and treatment recommended by an examining physician if a worker exhibits signs or symptoms associated with exposure to a hazardous material. A medical consultation must be offered to any employee potentially exposed through a spill, leak, or explosion of a hazardous chemical. The employer must provide information about the hazardous chemical, conditions under which the exposure occurred, and a description of symptoms experienced by the worker. The employer must obtain information on any medical condition that might pose an increased risk and a statement that the employee was informed of the results of the medical examination/consultation (Table 4.22).

Employers must develop criteria for determining and implementing control measures to ensure the use of engineering innovations, work practice controls, and personal protective equipment (PPE). Engineering controls should address proper ventilation, including the use of fume hoods, glove boxes, and other exhaust systems. Work practice controls may cover items such as restricting eating and drinking areas, prohibiting mouth pipetting, and performing work in a manner that minimizes exposure. Respiratory protection is to be used only as an interim measure or when engineering or work practice controls are not feasible. Use of respiratory equipment must comply with the requirements in 29 CFR 1910.134. Other PPE used in laboratories, as appropriate, include safety glasses, whole-body coverings, and gloves. Employers must include information on protective measures for work that involves carcinogens, reproductive toxins,

Table 4.22 OSHA Laboratory Standards' Training Topics

- Location of chemical hygiene plan, the chemical list, and SDSs
- Policies for handling, storing, and disposing of hazardous chemicals
- Elements of the hygiene plan and how it is implemented
- Hazards of the chemicals and the protective measures
- Specific procedures to protect workers
- Detection methods and observation guidelines

and acutely toxic substances. Establish a designated area with appropriate signs warning of the hazards associated with the substance. Provide information on safe and proper use of a fume hood or equivalent containment device. The Centers for Disease Control and Prevention (CDC) publishes bio-safety guidelines and requirements for research laboratories conducting operations with biohazards. The guidelines suggest using four levels of protections, depending on the hazard level of the laboratory operations.

Level 1 precautions provide for basic containment measures that depend on adhering to standard microbiological practices. This level does not prescribe special primary or secondary barriers, except for a sink for washing hands. Safety equipment and facilities must meet requirements for the appropriate education sessions taking place in teaching laboratories.

Level 2 precautions stress the use of secondary barriers and making waste decontamination available to reduce contamination. This level of protection is appropriate for tasks involving human blood, body fluids, or tissues. Primary hazards can include percutaneous events, mucous-membrane exposures, or ingestion of infectious materials. Use extreme precaution when working with contaminated needles or sharp instruments.

Level 3 precautions place more emphasis on primary and secondary barriers to protect personnel in contiguous areas. Ensure performance of all laboratory manipulations occur in a biological safety cabinet or other enclosed equipment. Secondary barriers include controlled access to the laboratory and a specialized ventilation system that minimizes the release of infectious aerosols. This level applies to clinical, diagnostic, teaching, research, and production facilities working with indigenous or exotic agents. Primary hazards to personnel working with these agents relate to automatic inoculation situations, ingestion, and exposure to infectious aerosols.

Level 4 precautions require complete isolation of aerosolized infectious materials primarily by working in a Class III biological safety cabinet or a full-body, air-supplied positive pressure suit. A Level 4 facility is generally a separate building or areas completely isolated within the complex. It requires specialized ventilation and waste-management systems. The primary hazards are respiratory exposure to infectious aerosols, mucous-membrane exposure to infectious droplets, and auto-inoculation. Level 4 practices would apply to safety equipment and facilities working with dangerous and exotic agents that pose a high life-threatening risk.

Infection Control

The Bloodborne Pathogens Standard and OSHA enforcement activities related to pandemic and tuberculosis exposures highlight the need for controlling biohazards in the workplace. Infection control planning no longer applies only to health-care and institutional settings. An effective infection control program should stress sound personal hygiene, individual responsibility, monitoring, and investigating infectious diseases with potentially harmful infectious exposures. The program should also stress providing care for work-related illnesses, identifying infection risks, instituting preventive measures, eliminating unnecessary procedures, and preventing infectious diseases. Implement a program to control important infection issues by developing a program using sound research and demographic considerations. Some professionals recommend the use of a hazard vulnerability analysis approach to guide the design process. CDC publishes guidelines, advisories, or recommendations that do not carry the force of law. The guidance offered by CDC gives infection control personnel the information necessary to make informed decisions. Keep exposed workers updated on the latest OSHA requirements and CDC developments with periodic in-service sessions. The continuous evaluation of care practices under the supervision of the infection control staff can help assure continued adherence to correct practices. Standard precautions consider hand washing as the first line of defense in preventing exposures to diseases, bloodborne pathogens, and infections. Standard precautions synthesize the major features of blood and body fluid precautions designed to reduce the risk of transmission of bloodborne pathogens.

Bacteria can adapt and survive in diverse environmental conditions. Bacteria normally exhibit one of three typical shapes: (1) rod, (2) round, or (3) spiral. Aerobic forms of bacteria function only in the presence of free or atmospheric oxygen. Anaerobic bacteria cannot grow in the presence of free oxygen but obtain oxygen from other compounds. Bacteria do not make their own food and must live in the presence of plant or other life. Methicillin-resistant *Staphylococcus aureus* (MRSA) is a bacterium that can live on the skin or in the nose of healthy people.

Viruses are smaller than bacteria and contain a chemical compound containing protein. They must infect a host to survive for long periods. Disinfectants destroy viruses very easily. Viruses depend on the host cell to reproduce. Outside a host cell, a virus exists as a protein coat that can be enclosed within a membrane. Although outside the cell, a virus remains metabolically inert. Viruses cause many diseases, including smallpox, colds, chickenpox, influenza, shingles, and hepatitis.

Many experts divide chemical germicides into three basic or general categories. Sterilizing agents were developed to eliminate all microbial life on objects or surfaces, including bacterial spores that can survive other germicides. Disinfectants can be classified as high, medium, or low level, depending on the strength required. They have the ability to destroy most microbial life, except for bacterial spores, on objects or surfaces. Antiseptics are used to inactivate or destroy organisms on skin or living tissue. The EPA oversees the manufacture, distribution, and use of disinfectants. Manufacturers must use preestablished test procedures to ensure product stability, determine toxicity to humans, and assess microbial activity. The EPA regulates disinfectants under the authority of the Federal Insecticide, Fungicide, and Rodenticide Act (FIFRA). The FDA regulates liquid-chemical sterilizing agents and high-level disinfectants such as hydrogen peroxide and peracetic acid under the authority of the 1976 Medical Devices Amendment to the Food, Drug, and Cosmetic Act. The FDA also regulates the chemical germicides if marketed for use on specific medical devices.

The OSHA Bloodborne Pathogens Standards requires workers to wash their hands immediately or as soon as feasible after removal of gloves or other PPE. Paragraph (vi) states that workers shall wash hands and any other skin with soap and water, or flush mucous membranes with water, immediately or as soon as feasible following contact of such body areas with blood or other potentially infectious materials. Workers removing gloves after exposure to blood or other potentially infectious materials must wash hands in appropriate soap and running water. Workers with no access to a readily available sink after an exposure may decontaminate hands with a hand cleanser or towelette. However, workers must wash hands with soap and running water as soon as feasible.

The OSHA Bloodborne Pathogens Standard sets forth requirements for employers to implement an exposure-control plan for each worksite with potential employee exposure. The mandated plan must describe how an employer will use engineering, work-practice controls, and wear personal protective clothing to protect workers. The plan also details training, medical surveillance, hepatitis B vaccinations, and signs and labeling requirements.

The plan explains the employer's exposure determination procedures. Employers must determine all job classifications in which employees will be exposed or may occasionally be exposed without regard to the use of PPE. The employer shall take appropriate preventative measures against occupational exposure. These include engineering controls and work-practice controls.

Engineering controls include hoods, puncture-resistant sharps containers, mechanical pipette devices, and others devices to permanently remove the hazard or help isolate the worker from exposure. Work-practice controls include hand-washing policies, sharps handling procedures, and proper waste and disposal techniques. Employers shall provide PPE to eliminate or minimize the risk of infectious material entering employees' bodies.

The employer may not institute a program in which the employees pay the original cost of the vaccine and the company reimburses them if employed for a specified period of time. OSHA requires employers to offer HBV vaccination during employee scheduled work hours.

OSHA requires all employees with occupational exposure to receive initial and annual training on the hazards associated with exposure to bloodborne pathogens. The training session must cover the topics listed in the standard. Training requires site-specific information to be presented. The annual retraining for these employees must be provided within 1 year of their original training. Refresher training must cover topics listed in the standard to the extent needed and must emphasize new information or procedures. Employers must train part-time, temporary employees, and workers referred to as agency or per diem employees.

Private and Industrial Security

The history of the private security dates back to the 1850s, when Allan Pinkerton founded his national detective agency in Chicago. The agency quickly became the largest private security company in the United States. The agency's work for the railroads helped build an international reputation for the company. The Brinks Company came into existence in 1889 for the purpose of protecting payrolls and property. William J. Burns formed a private detective agency in 1909 that became American Banking Association's investigative unit. During World War II, thousands of military personnel served in law enforcement and intelligence operations to protect against espionage and sabotage. The Cold War, which began in the 1950s, created the need for background investigations and security clearances, which provided civilian jobs for these highly trained individuals. Today, the private security industry protects many of the nation's institutions and critical infrastructure systems. They also protect intellectual property and sensitive corporate information. Many companies also rely heavily on private security for a wide range of functions, including protecting employees and property, conducting investigations, performing preemployment screening, providing information technology security, and many other functions. Since September 11, 2001, private security grew to become a significant force in public security and safety. A security officer can work in a variety of positions with varying responsibilities or functions.

The American Society of Industrial Security (ASIS), founded in 1955, promotes the professionalism of private security. ASIS develops educational programs and materials that address a broad range of security topics. ASIS also conducts an annual professional development conference for its more than 30,000 members. Under the broad definition commonly used today, the term *private security* can represent a wide range of organizations, including corporate security, security guard companies, armored car businesses, investigative services, and many others. Personnel hired by these companies can serve in armed or unarmed positions. According to ASIS, proprietary security refers to any organization or department of that organization that provides full-time security officers solely for itself.

Private security officer selection and training criteria vary from state to state, ranging from comprehensive training requirements for every private security officer to little or no training for private security officers. Effective security today requires workers to understand all aspects of a facility's security system for assessing and containing potential threats. Security officers must understand emergency

Table 4.23   ASIS Defined "Core Elements" of Private Security Education

1. Physical security
2. Personnel security
3. Information-systems security
4. Investigations
5. Loss prevention
6. Risk management
7. Legal aspects
8. Emergency and contingency planning
9. Fire protection
10. Crisis management
11. Disaster management
12. Counterterrorism
13. Competitive intelligence
14. Executive protection
15. Workplace violence
16. Crime prevention
17. Crime prevention through environmental design
18. Security architecture and engineering

procedures, hazard control management principles, accident-prevention concepts, and investigation techniques. They must also work closely and effectively with public safety personnel, including first responders. The ability to protect the nation's critical infrastructure and contribution to Homeland Security efforts depends largely on the competence of private security officers (Table 4.23).

Fleet and Vehicle Safety

Transportation affects nearly every enterprise and organization. Companies with fleet or driver safety functions must make driver selection and qualification critical to success. Ensure each applicant who will operate a motor vehicle completes a formal application and signs a release to permit the company to obtain their formal driving records from the Department of Motor Vehicles. Conduct substance-abuse testing, check all references, and provide training sessions. Many insurance carriers provide training on fleet safety. They normally "train the trainer," who then will conduct sessions for drivers. Some safety councils provide training sessions and even specialized training classes. The effective supervision of drivers poses the greatest challenge to the fleet manager. Some organizations use "how's my driving" bumper stickers, encouraging reporting of erratic driving. The information reported can determine trends or document multiple reports on the same driver. The driver then can receive counseling or face discipline or termination. The psychological component to this approach usually gets each driver's attention. Other technologies permit employers to monitor driver safety and even location. Effective driver policies can guide staff decision making, establish conduct standards, and help with operational consistency. Publish driver policies in writing, and ensure their wide dissemination.

Table 4.24 Key Elements of a Fleet Safety Plan

- Management involvement and safety-policy development
- Safety supervision and selection of safe drivers
- Safety training and employee involvement in the process
- Vehicle safety, including maintenance, reporting, investigation, and analysis
- Safety promotion

NIOSH established the Center for Motor Vehicle Safety (NCMVS) in December 2010 to promote research related to preventing motor vehicle crashes. Traffic accidents remain the leading cause of occupational fatalities in the United States. NCMVS's initiatives address road safety for workers across all industries and occupations, based on rigorous assessment of research needs. Research will target all potential risk factors for work-related motor vehicle crashes. NIOSH will focus on use of occupant restraints, driver fatigue, vehicle design, work organizational factors, and employer policies. Research begins with collecting injury data that can lead to identification of risk factors. NIOSH hopes to develop injury prevention strategies and transfer the information into workplaces. The National Safety Council publishes the "Fleet Safety Manual," which is an excellent resource for promoting organizational motor vehicle safety (Table 4.24).

Safe Practices for Motor Vehicle Operations (ANSI/ASSE Z15.1 Standard)

This standard sets forth practices for the safe operation of motor vehicles owned or operated by organizations, including definitions, management, leadership, administration, operational environment, driver considerations, vehicle considerations, and incident reporting/analysis. The standard provides organizations with a guidance document to assist with the development of policies, procedures, and processes necessary to control risks related to the operation of motor vehicles. The standard applies to licensed motor vehicles designed to be operated primarily on public roads. The standard applies to the operation of organization-owned or leased vehicles. The standard also applies to persons working on behalf of the organization whose job performance requires the use of a motor vehicle. The standard does not apply to unlicensed equipment or off-road recreational vehicles. The standard sets forth practices for the safe operation of motor vehicles owned or operated by organizations. The standard now contains the definition of an "organizational vehicle." The revised standard connects the importance of the Z-10 Occupational Safety and Health Management Systems Standard to the Z-15 requirements. Motor vehicle practices and operations play a vital role in the effectiveness of any overall safety-and-health management function. The newly revised standard places added emphasis on restraint systems, impaired driving, aggressive driving, distracted driving, journey management, and fatigue management. The revised standard also places an increased emphasis on the vehicle acquisition, inspection, and maintenance. The standard recommends that organizations develop written motor vehicle safety policies to meet organizational needs.

Review Exercises

1. Define the concept known as static coefficient of friction.
2. Describe the purpose of the EPA-developed NRR.
3. List the three types of equipment necessary for an effective HVAC system.

4. Explain the difference between SBS and building-related illness.
5. List the four OSHA-defined conditions that would require a confined space to be designated as "permit required."
6. List at least four elements of a design-related weakness.
7. List at least four elements of an operational-related weakness.
8. What is the fundamental purpose of effective safety engineering?
9. Define the terms *safety critical, normative safety,* and *system reliability.*
10. Briefly describe the following processes and their purpose:
    a. Hazard and operability (HAZOP)
    b. Failure mode, effects, and criticality analysis (FMECA)
    c. Fault-tree analysis (FTA)
    d. Instructional systems design (ISD)
11. List and describe the five system-related functions.
12. Define the practice of industrial hygiene.
13. Define the term macroergonomics.
14. List at least five factors impairing human performance.
15. Define an industrial robot.
16. Define product safety.
17. List the five basic elements of a fleet safety plan.

Chapter 5

Government, Consensus, and Voluntary Organizations

Introduction

Hazard control personnel must know about the many governmental agencies involved in promoting and regulating the various aspects of safety and health. Agencies such as Occupational Safety and Health Administration (OSHA), Environmental Protection Agency (EPA), Nuclear Regulatory Commission (NRC), Food and Drug Administration (FDA), and Department of Transportation (DOT) publish standards and hold enforcement powers. Other federal agencies such as the National Institute of Occupational Safety and Health (NIOSH) and Centers for Disease Control and Prevention (CDC) do not enforce standards but provide guidance, publications, and information relevant to hazard control–related topics. Hazard control personnel should also understand the role that voluntary or consensus standards organizations make in supporting effective hazard control practice. Finally, hazard control personnel should understand the professional opportunities that membership groups and professional associations provide. To learn more about a particular agency, organization, society, or association, access their individual websites.

Federal Register

The US Code serves as the "official document," containing general and permanent laws of the United States. The US Code compiles laws in force from 1789 to the present day. As prima facie, or presumed to be law, it does not include repealed or expired acts. The Federal Register Act became law on July 26, 1935, and established a uniform system for handling governmental agency regulations. An amendment in 1937 established permanent codification through a numerical arrangement of rules in the Code of Federal Regulations (CFR). The act required agencies to make documents available for public inspection through publication of documents in the Federal Register (FR). Congress passed the Administrative Procedure Act in 1945. That act gives the public an opportunity to participate in the rule making process by commenting on proposed standards. The act provides for the publication of federal agency statements about organizational and procedural rules. Informing the public about proposed rule making actions remains the primary purpose of the FR. The National Archives and Record Service publish the FR each workday. The publication provides the public access to federal regulations, proposed standards, and other legal notices. Most federal agencies publish their regulatory agendas and strategic plans in the FR. The pages contained in the FR are numbered sequentially throughout the year, beginning January 1 and ending December 31.

Code of Federal Regulations

The National Archives and Record Services also publish the CFR annually in paperback volumes. The CFR contains a compilation of rules previously published in the FR. This results in a codification of the rules, which take the force of administrative law. The government divides the CFR into 50 titles representing areas of federal regulation. Currency of the CFR happens daily, with updates published daily in FR. The CFR contains 50 divisions, which are revised annually (Table 5.1).

Occupational Safety and Health Administration

The Occupational Safety and Health Act of 1970, also known as the OSH Act, established OSHA and the NIOSH. Employee and employer coverage comes from federal OSHA jurisdiction or by a state plan approved by OSHA. The act defines an employer as any person engaged in a business affecting commerce with employees, but does not include any state or political subdivision of a state.

Therefore, the act applies to employers and employees in such varied fields as manufacturing, construction, longshoring, agriculture, law and medicine, charity and disaster relief, organized labor, and private education. The act establishes a separate program for federal government employees and extends coverage to state and local government employees only through the states with OSHA-approved plans. The act does not cover self-employed persons or farms that employ only immediate family in the operation. The act does not address working conditions of other federal agencies that operate under the authority of other federal laws regulating worker safety.

The act does not apply to employees of state and local governments, unless they're located in a state that operates an OSHA-approved state plan. The act assigns OSHA the functions of setting standards and conducting inspections to ensure that employers provide safe and healthful workplaces. Refer to 29 CFR 1975 for additional information about determining what organizations must comply with OSHA standards. OSHA standards may require that employers adopt certain

Table 5.1 Key Hazard Control–Related CFR Titles

- Title 10, Energy and Radiation (NRC)
- Title 21, Food and Drugs (FDA)
- Title 29, Labor (OSHA)
- Title 40, Environmental Protection (EPA)
- Title 42, Health and Human Services (NIOSH, CMS)
- Title 44, Emergency Management (FEMA)
- Title 49, Transportation (DOT)

practices, means, methods, or processes reasonably necessary and appropriate to protect workers on the job. Employers must become familiar with the standards applicable to their establishments and eliminate hazards. Compliance with standards may require implementing engineering or administrative controls. OSHA requires employee training on the use of personal protective equipment (PPE) when other controls provide inadequate protection. Employees must also comply with all rules and regulations that apply to their own actions and conduct. In areas where no OSHA standard exists for a specific standard, employers are responsible for complying with the act's "general duty" clause. Some standards with similar requirements for all sectors of industry include those that address access to medical and exposure records, PPE, and hazard communication (Tables 5.2 and 5.3).

Every workplace covered by the act must display the OSHA Poster (Publication 3165) or the poster for the state equivalent plan. Employers should refer to 29 CFR 1975 for information about which entities fall under OSHA jurisdiction. The OSHA Poster explains worker rights to a safe workplace and how to file a complaint. Employers must place the poster in a highly visible location. OSHA compliance officers can conduct "programmed inspections" using OSHA-established selection criteria. These criteria can include injury rates, fatality rates, exposure to toxic substances, or a high number of lost workdays for certain industries. A "non-programmed inspection" can occur when an employee makes a formal complaint to OSHA regarding a possible unsafe working condition or imminent danger at the workplace. If the inspector arrives without a warrant, employers may deny access to the facility, based on the Fourth Amendment.

OSHA regulations require employers to report deaths on the job within 8 hours of becoming aware of the fatality. Imminent danger situations receive top priority. OSHA defines imminent danger as any condition with reasonable certainty that a danger exists. The danger or hazardous situation could result in immediate death or serious physical harm. When confirming that an imminent danger situation exists, the compliance officer will ask the employer to voluntarily abate

Table 5.2   General Duties of Employers & Employees (OSH ACT)

- 5(a) (1)—Each employer shall furnish to each of his employees employment and a place of employment which are free from recognized hazards that are causing or are likely to cause death or serious physical harm to his employees.

- 5(a) (2)—Each employer shall comply with occupational safety and health standards promulgated under this Act.

- 5(b) Each employee shall comply with occupational safety and health standards and all rules, regulations, and orders issued pursuant to this Act which are applicable to his own actions and conduct.

Table 5.3   Types of OSHA Standards

- General Industry Standards, 29 CFR 1910
- Occupational Safety and Health Standards for Shipyard Employment, 29 CFR 1915
- Marine Terminals Standards, 29 CFR 1917
- Longshoring Standards, 29 CFR 1918
- Construction Standards, 29 CFR 1926

the hazard or remove endangered employees from exposure. Should the employer refuse, OSHA will take legal action in federal court to correct the situation. The second inspection priority relates to investigation of fatalities and catastrophes resulting in hospitalization of three or more employees (Table 5.4).

Proposed penalties can vary depending on the statutory factors, such as employer size, gravity of violation, good faith of the employer, and the history of previous violations. OSHA penalties do not correspond to or reflect the value of a worker's life or the cost of injury or illness. The act contains a provision that permits OSHA to document when an employer willfully violates an OSHA standard.

If the violation caused the death of a worker, OSHA can refer the matter to the Department of Justice (DOJ) for criminal prosecution. Criminal referral remains a strong enforcement tool that OSHA can use in serious cases. Each criminal violation allegation to include willfulness must go to a jury and meet the "beyond a reasonable doubt" threshold. An "other than serious violation" has a direct relationship to job safety and health, but probably would not cause death or serious physical harm.

A serious violation is issued when there is substantial probability that death or serious physical harm could result, and that the employer knew, or should have known, of the hazard. Imminent danger situations are also cited and penalized as serious violations. A willful violation is one that the employer intentionally and knowingly commits. The employer either knows that the operation constitutes a violation, or is aware that a hazardous condition exists and made no reasonable effort to eliminate it. A repeat violation can be issued to address any standard regulation, rule, or order where, on reinspection, another violation of the same previously cited section is found. Falsifying records, reports, or applications can bring a fine or jail time upon conviction. Assaulting a compliance officer, or otherwise resisting, opposing, intimidating, or interfering with a compliance officer in the performance of his or her duties is a criminal offense.

OSHA Alliances enable organizations committed to workplace safety and health to collaborate with OSHA to prevent injuries and illnesses in the workplace. OSHA and its allies work together to reach out to, educate, and lead the nation's employers and their employees in improving and advancing workplace safety and health. OSHA makes alliances with trade or professional organizations, businesses, labor organizations, educational institutions, and government agencies. In some cases, organizations may build on existing relationships with OSHA through

Table 5.4 OSHA Inspection Priorities

- Imminent Danger Situations—hazards that could cause death or serious physical harm.
- Fatalities and Catastrophes—incidents that involve a death or the hospitalization of three or more employees.
- Complaints—inspections into allegations of hazards present or violations of standards.
- Referrals—hazard information obtained from other federal, state, or local agencies, individuals, organizations, or the media will receive consideration for inspection.
- Follow-up Inspections—checks for abatement of violations cited during previous inspections and are also conducted by the agency in certain other circumstances.
- Planned or Programmed Investigations—inspections aimed at specific high-hazard industries and individual workplaces with high rates of injuries and illnesses.

other cooperative programs. The OSH Act provides for a wide range of substantive and procedural rights for employees and representatives of employees. Section 11(c) of the act prohibits any person from discharging or in any manner retaliating against any employee because the employee has exercised rights under the act.

These rights include complaining to OSHA and seeking an inspection, participating in an inspection, and participating or testifying in any proceeding related to an inspection. OSHA also administers the whistle-blowing provisions of other statutes, protecting employees who report violations of various airline, commercial motor carrier, consumer product, environmental, financial reform, health-care reform, nuclear, pipeline, public transportation agency, railroad, maritime, and securities laws. A person filing a complaint of discrimination or retaliation must show that he or she is engaged in protected activity, the employer knew about that activity, the employer subjected him or her to an adverse action, and the protected activity contributed to the adverse action. OSHA generally defines adverse action as any action that would dissuade a reasonable employee from engaging in protected activity. Depending upon the circumstances of the case, adverse action can include the following: (1) firing or laying-off, (2) blacklisting, (3) demoting, (4) denying overtime or promotion, (5) disciplining, (6) denial of benefits, (7) failure to hire or rehire, (8) intimidation, (9) making threats, (10) reassignment affecting prospects for promotion, and (11) reducing pay or hours.

The act requires the secretary to develop and maintain an effective program of collection, compilation, and analysis of occupational safety and health statistics. 29 CFR 1904, Recording and Reporting Occupational Injuries and Illnesses, requires employers to record information on the occurrence of injuries and illnesses in their workplaces. The employer must record work-related injuries and illnesses that meet one or more of certain recording criteria. OSHA rules found in 29 CFR part 1904 require all employers with 11 or more employees to keep OSHA injury and illness records, unless classified in a specific low-hazard industry. Employers with 10 or fewer employees must keep OSHA records, if OSHA or the Bureau of Labor Statistics informs them in writing that they must keep records. Employers may use the OSHA 301 or an equivalent form that documents the same information. Some state workers compensation, insurance, or other reports may be acceptable substitutes, as long as they provide the same information as the OSHA 301. The OSHA Log of Work-Related Injuries and Illnesses (Form 300) serves as a means to document and classify work-related injuries and illnesses. The log also documents the extent and severity of each case.

The OSHA summary (Form 300A) reflects the totals for the year in each category. Employers must post the summary from February 1 through April 30 of each year. Post in a visible location to inform employees about the injuries and illnesses that occurred in their workplace. Employers must keep a log for each establishment or site. If you have more than one establishment, you must keep a separate log and summary for each physical location that is expected to be in operation for 1 year or longer. All logs must be kept for 5 years.

Occupational Safety and Health Review Commission

OSHRC, an independent agency, decides contests of citations or penalties resulting from OSHA inspections. The review commission functions as an administrative court, with established procedures for conducting hearings, receiving evidence, and rendering of decisions by its administrative law judges (ALJs). OSHRC membership consists of a three-member board appointed by the president and confirmed by the Senate. It is an independent agency of the executive branch. The commission may conduct investigations. They can uphold, change, or dismiss OSHA findings. Some cases are determined by an administrative law judge, but the three-member board has the final rule. Rulings by OSHRC can be reviewed further by the federal appeals courts. Rules of procedures are found in 29 CFR 2200. The burden of proof rests with the government attorney.

Environmental Protection Agency

The Environmental Protection Agency (EPA) works to protect human health and environment by promulgating enforceable compliance standards based on environmental laws passed by Congress. On July 9, 1970, President Richard Nixon transmitted Reorganization Plan No. 3 to Congress by executive order. This action created the EPA as a single and independent agency, taking over responsibilities from a number of smaller agencies. Before the creation of the EPA, the federal government lacked the structure to comprehensively regulate environmental pollutants. On December 2, 1970, the EPA began operation. The president appoints the administrator who must receive congressional approval. The administrator functions at the cabinet rank. Many of the EPA's rules define which substances can be hazardous to human health or pose a threat to the environment. The EPA or state-approved agencies provide guidance for handling hazardous materials, regulate the operation of waste disposal sites, and establish procedures for dealing with environmental incidents such as leaks or spills. It publishes informative guides to assist risk managers in understanding and complying with a number of environmental laws and regulations. Refer to 40 CFR to access U.S. environmental laws.

The Resource Conservation and Recovery Act (RCRA) serves as the primary law governing the disposal of solid and hazardous waste. Congress passed RCRA on October 21, 1976, to address the increasing problems of a volume of municipal and industrial waste. RCRA amended the Solid Waste Disposal Act of 1965 to protect human health and the environment from potential hazards of waste disposal. The legislation also sought to conserve energy and natural resources, reduce amounts of generated wastes, and ensure management of wastes in an environmentally sound manner.

The solid waste program, under RCRA Subtitle D, encouraged states to develop comprehensive plans to manage nonhazardous industrial solid waste and municipal solid waste, sets criteria for municipal solid waste landfills and other solid waste disposal facilities, and prohibits the open dumping of solid waste. The hazardous waste program, under RCRA Subtitle C, establishes a system for controlling hazardous waste from the time it is generated until its ultimate disposal.

The underground storage tank (UST) program, under RCRA Subtitle I, regulates USTs containing hazardous substances and petroleum products. RCRA banned open dumping of waste and encouraged source reduction and recycling, thereby promoting safe disposal of municipal waste. RCRA mandates strict controls over the treatment, storage, and disposal of hazardous waste. The first RCRA regulations, "Hazardous Waste and Consolidated Permit Regulations," published in the Federal Register on May 19, 1980, established the "cradle to grave" approach to hazardous waste management that still exists. Congress amended RCRA in November 1984 with the passing of the Federal Hazardous and Solid Waste Amendments (HSWA). These amendments to RCRA required phasing out land disposal of hazardous waste. Some of the other mandates of this strict law include increased enforcement authority for EPA, more stringent hazardous waste management standards, and a comprehensive UST program. RCRA has been amended on two occasions since HSWA. The first occurred with the federal Facility Compliance Act of 1992, which strengthened enforcement of RCRA at federal facilities. The Land Disposal Program Flexibility Act of 1996 provided regulatory flexibility for land disposal of certain wastes. RCRA focuses only on active and future facilities and does not address abandoned or historical sites currently managed under the Comprehensive Environmental Response, Compensation, and Liability Act (CERCLA), commonly known as the Superfund. Hazardous waste generators must identify and label all wastes as well as notifying the EPA about hazardous waste operations. Sites must maintain secure storage areas, keep appropriate records, and train all waste handlers. Facilities shipping waste must use permitted treatment, storage, and disposal facilities.

The CERCLA provided the government authority to remedy the past mistakes in hazardous waste management. RCRA strives to help avoid those types of mistakes through proper management of waste activities. CERCLA, as originally enacted in 1980, authorized a 5-year program by the federal government to perform cleanup activities of contaminated sites where release of hazardous substances pose a serious threat to human health, welfare, or the environment. CERCLA held the parties responsible for the releases to fund the cleanup actions.

The Superfund Amendments and Reauthorization Act of 1986 (SARA) established new standards and schedules for site cleanup. SARA also required informing the public of risks from hazardous substances in the community and preparing communities for hazardous substance emergencies. SARA specified new requirements for state and local governments. The Emergency Planning and Community Right-to-Know Act specifically require states to establish a State Emergency Response Commission (SERC). The SERC must designate emergency planning districts within the state. The SERC appoints a Local Emergency Planning Committee (LEPC) for each district. SERC and LEPC responsibilities include implementing various planning provisions of Title III and serving as points of contact for the community right-to-know reporting requirements.

The Clean Air Act (CAA) limits the emission of pollutants into the atmosphere and protects human health and the environment from the effects of airborne pollution. The EPA established National Ambient Air Quality Standards (NAAQS) for several substances. The NAAQS provide the public some protection from toxic air pollutants. Primary responsibility for meeting the requirements of the CAA rests with each state. States must submit plans for achieving NAAQS compliance. Under section 112 of the CAA, the EPA has the authority to designate hazardous air pollutants and set national emission standards for hazardous air pollutants. Common air pollutants include the following: (1) ozone, (2) nitrogen dioxide, (3) carbon monoxide, (4) particulate matter, and (5) sulfur dioxide.

The Clean Water Act of 1977 strengthened and renamed the Federal Water Pollution Control Act of 1972. The act includes several major provisions related to the establishment of the National Pollutant Discharge Elimination Systems (NPDES) to regulate discharges into the nation's waterways. Any direct discharges into surface water require a NPDES permit. An indirect discharge means that the waste is first sent to a publicly owned treatment works and then discharged pursuant to a permit. The act requires regulation of certain industrial and municipal storm-water discharges through the NPDES permit system.

The Federal Insecticide, Fungicide, and Rodenticide Act (FIFRA) of 1947 gave the Department of Agriculture authority to regulate pesticides. The EPA became responsible for the act in 1970, and a 1972 amendment outlined new provisions that provided for the protection of public health and the environment. FIFRA controls risks of pesticides through the EPA registration system. Each new pesticide must achieve registration before going to market. The EPA can refuse to register a pesticide or can limit use if evidence indicates a threat to humans and the environment. The EPA must also approve and register all general disinfectants.

Toxic Substances Control Act (TSCA) enacted in 1976 helps control the risk of substances not regulated as drugs, food additives, cosmetics, or pesticides. Under this law, the EPA regulates the manufacture, use, and distribution of chemical substances. The TSCA mandates EPA notification before the manufacture of any new chemical substance. The EPA ensures that all chemicals undergo testing to determine risks to humans. The TSCA also allows the EPA to regulate polychlorinated biphenyls (PCBs) under 40 CFR 761.

Department of Transportation

DOT serves as the primary agency with the responsibility for shaping and administering policies and programs to protect/enhance the safety, adequacy, and efficiency of the transportation system and services. The DOT oversees the following agencies: Federal Aviation Administration (FAA), Federal Highway Administration (FHWA), Federal Motor Carrier Safety Administration (FMCSA), Federal Railroad Administration (FRA), and National Highway Traffic Safety Administration (NHTSA).

The FMCSA became a separate administration within DOT on January 1, 2000. The FMCSA works to reduce crashes, injuries, fatalities, and property loss involving large trucks and buses by regulating the workers involved in these industries. The administration develops and enforces data-driven regulations that balance motor carrier safety with industry efficiency. The administration issues motor-carrier numbers to "For Hire Interstate Motor Carriers" who transport passengers, property, and hazardous materials. The administration also enforces hazardous material regulations (HMRs) designed to ensure the safe and secure transportation of hazardous materials. These rules address the classification of hazardous materials, proper packaging, employee training, hazard communication, and operational requirements.

The Pipeline and Hazardous Materials Safety Administration (PHMSA) oversees pipeline and hazardous materials transportation safety. The administration oversees the nation's pipeline infrastructure, which accounts for 64% of energy commodities consumed in the United States. The Office of Pipeline Safety (OPS) functions as the safety authority for the nation's 2.3 million miles of natural gas and hazardous liquid pipelines. The office administers a national regulatory program to ensure the safe pipeline transportation of natural gas, liquefied natural gas, and hazardous liquids by pipeline. The Accountable Pipeline Safety and Partnership Act of 1996 requires that OPS adopt rules requiring interstate gas pipeline operators to provide maps of their facilities to the governing body of each municipality in which a pipeline traverses.

The Highway Safety Act of 1970 established the NHTSA. In 1972, the Motor Vehicle Information and Cost Savings Act expanded NHTSA scope to include consumer information programs. NHTSA is charged with writing and enforcing safety, theft-resistance, and fuel-economy standards for motor vehicles. NHTSA also licenses vehicle manufacturers and importers, allows or blocks the import of vehicles and safety-regulated vehicle parts, and administers the vehicle identification number system. It develops the anthropomorphic dummies used in safety testing and establishes the test protocols, and provides vehicle insurance cost information. The Fatality Analysis Reporting System has become a resource for traffic safety research. The Federal Motor Vehicle Safety Standards are contained in 49 CFR 571.

The FAA, formally created by the Federal Aviation Act of 1958, actually began operation in 1926 with the passing of the Air Commerce Act. The FAA engages in a variety of activities to fulfill its responsibilities, including safety regulation. The FAA issues and enforces rules, regulations, and minimum standards relating to the manufacture, operation, and maintenance of aircraft. The FAA also rates and certifies people working on aircraft, including medical personnel, and certifies airports that serve air carriers. The agency performs flight inspections of air-navigation facilities in the United States and, as required, abroad.

FAA employs hundreds of safety inspectors, who oversee all planes operated by scheduled airlines and more than 200,000 additional aircraft. It also inspects thousands of repair stations and hundreds of pilot training schools. However, the nation's Air Traffic Control (ATC) remains the FAA's most visible mission. The FAA also provides airport construction grants and regulates many aspects of aviation, including airport safety and security. The agency regulates the design, manufacture, and maintenance of aircraft, including spare parts.

The FRA promulgates and enforces rail safety regulations, administers railroad assistance programs, conducts research, and coordinates innovation to support railroad safety and national rail transportation policy. The administration also provides for the rehabilitation of Northeast Corridor rail passenger service and consolidates government support of rail transportation activities. The Office of Railroad Safety promotes and regulates safety throughout the country's rail industry. It employs more than 415 federal safety inspectors operating out of eight regional offices. FRA safety inspectors specialize in five disciplines. They also address many grade crossings and trespass-prevention initiatives, including track, signal and train control, motive power and equipment, operating practices, hazardous materials, and highway rail grade crossing hazards.

The Federal Transit Administration (FTA) helps keep public transportation systems moving and assists transit agencies. Public transportation includes buses, subways, light rail, commuter rail, monorail, passenger ferries, trolleys, inclined railways, commuter vanpools, and people movers. The FTA provides financial assistance to develop new transit systems and improve, maintain, and operate existing systems. The FTA oversees grants to states and local transit providers and ensures grantees manage their programs in accordance with federal requirements. The FTA now publishes a safety guidebook with the objective of providing resource information for transit agencies and the FTA regarding the development and implementation of Safety Management Systems (SMS) and Safety Performance Measurement Systems (SPMS).

The Maritime Administration (MARAD) maintains the National Defense Reserve Fleet (NDRF) as a ready source of ships for use during national emergencies. Some consider this traditional role as the nation's fourth arm of defense vital to supporting military operations when needed. In 1961, the Federal Maritime Commission assumed the Federal Maritime Board regulatory functions. During August 1981, MARAD came under control of DOT. MARAD administers financial programs to develop, promote, and operate the Maritime Service and the US Merchant Marines. MARAD maintains equipment, shipyard facilities, and reserve fleets of government-owned ships essential for national defense. The MARAD administrator can assume residual powers as the director of the National Shipping Authority to organize and direct emergency merchant marine operations.

The FHWA, established in 1966, provides funding and technical support for the construction and preservation of highways. The FHWA budget provides support for local and state governments, as well as national parks, national forests, Indian lands, and other federally owned land. The organization receives funding from fuel and motor vehicle excise taxes. FHWA oversees federal funds used for constructing and maintaining the National Highway System. FHWA performs research in the areas of automobile safety, congestion, highway materials, and construction methods. FHWA also publishes the *Manual on Uniform Traffic Control Devices* (MUTCD), which is used by most highway agencies in the United States. The mission of FHWA Office of Safety remains a simple one: reduce highway fatalities by making roads safer. The agency uses a data-driven, systematic approach, and addresses all "4Es" of safety: engineering, education, enforcement, and emergency medical services. FHWA strives to provide decision makers with important information, tools, and resources that will improve the safety performance of roadways.

Mine Safety and Health Administration

This administration, an agency of the Department of Labor, develops and enforces safety and health rules applying to all US mines. They assist mine operators with special compliance problems and issues. The administration also provides technical, educational, and other types of assistance that support mine safety. The federal Mine Safety and Health Act (MSHA) of 1977 addresses specific MSHA responsibilities. The act applies to all mining and mineral processing operations in the US regardless of size, number of employees, or methods of extraction. MSHA accomplishes the enforcement of safety and health rules through two functional entities. The Coal Mine Safety and Health "activity" conducts mine inspections, investigations, and training programs through 11 district offices in the nation's coal mining regions. The Metal and Nonmetal Mine Safety and Health "activity" administers its programs for all noncoal mines through six district offices in mining areas throughout the United States. MSHA inspectors must issue a citation for each violation of a health or safety standard documented. The penalties may range up to $220,000 per violation. MSHA considers violations that could likely cause reasonably serious injury as "significant and substantial" violations. Assessment of these violations considers the following six factors: (1) history of previous violations, (2) size of the operator's business, (3) any negligence by the operator, (4) gravity of the violation, (5) the operator's good faith in trying to correct the violation promptly, and (6) effect of the penalty on the operator's ability to stay in business. MSHA investigates all complaints of discrimination and interference. When MSHA can find evidence of discrimination or interference, the Labor Department can take the miner's case before the Federal Mine Safety and Health Review Commission. They can seek appropriate relief, including back pay, civil penalties, and other remedies. The Mine Act provides for criminal sanctions against mine operators who willfully violate safety and health standards. If no resolution occurs, the operator may seek a hearing before an administrative-law judge with the independent Federal Mine Safety and Health Review Commission.

MSHA also identified training as an important tool for preventing accidents and avoiding unsafe or unhealthful working conditions. MSHA requires that each US mine operator establish an approved plan for miner training. This plan must provide 40 hours of basic safety and health training for new miners without any underground mining experience before beginning underground work. The plan must include 24 hours of basic safety and health training for new miners who have no surface mining experience before beginning work. MSHA requires eight hours of annual refresher safety and health training for all miners. Operators must provide safety-related task training to all miners assigned to new jobs.

Nuclear Regulatory Commission

Established as an independent agency by the Energy Reorganization Act of 1974, NRC regulates civilian use of nuclear materials. NRC operates from a Rockville, Maryland, headquarters and four regional offices. Headed by a five-member commission, the NRC regulates by-product, source, and special nuclear materials to ensure adequate public health and safety, common defense and security, and to protect the environment. The NRC or state governments under agreement with the agency can issue licenses for other uses of radioactive materials. The NRC also enforces standards for the departments of nuclear medicine in health-care facilities. Some states enter into NRC agreements to assume regulatory responsibilities. NRC issues 5-year licenses to qualified health-care organizations that follow prescribed safety precautions and standards. Types of regulated facilities include (1) nuclear power plants, (2) departments of nuclear medicine at hospitals, (3) academic activities at educational institutions, (4) research work in scientific organizations, and (5) industrial applications such as gauges and testing equipment.

Department of Health and Human Services

Department of Health and Human Services (DHHS) holds the responsibility of protecting the health of all Americans. DHHS represents almost a quarter of all federal outlays. It also administers more grant dollars than all other federal agencies combined. The DHHS Medicare program is the nation's largest health insurer and processes more than a billion claims each year. DHHS works closely with state and local governments. It also provides funding to many state and local agencies. The Centers for Disease Control (CDC), located in Atlanta, Georgia, is an agency of the DHHS. It works to protect the health of the American people by tracking, monitoring, preventing, and researching disease. It is also responsible for surveillance and investigation of infectious disease in health-care facilities. The CDC conducts research and publishes results in its *Morbidity and Mortality Weekly Report*. This weekly publication provides health-care facilities with timely information on topics such as infection control, isolation procedures, blood-borne pathogens, tuberculosis management, infectious waste disposal recommendations, and how to protect workers. CDC performs many of the administrative functions for both the NIOSH and the Agency for Toxic Substances and Disease Registry (ATSDR). CDC seeks to collaborate to create the expertise, information, and tools that people and communities need to protect their health—through health promotion; prevention of disease, injury, and disability; and preparedness for new health threats.

The NIOSH, created by the OSH Act, conducts research and makes recommendations to help prevent work-related injury and illness. NIOSH falls under administrative control of CDC. NIOSH operates research facilities in Cincinnati, Pittsburgh, Spokane, Atlanta, and Morgantown, West Virginia. The NIOSH Health Hazard Evaluation (HHE) function responds to requests for workplace evaluations from employers, employees and their representatives, and other agencies. Through the HHE program, NIOSH identifies current hazards and recommends practical, scientifically valid solutions for reducing exposures and preventing disease, injury, and disability. The National Personal Protective Technology Laboratory (NPPTL) located in Pittsburgh, Pennsylvania, provides research for the prevention of injury and illness among workers who must rely on personal protective equipment (PPE), including respirators, gloves, and hard hats. The strategic research program will ensure that the development of new PPE will meet real needs as work settings, technologies, and worker populations change and new threats emerge. NIOSH also supports training of occupational safety and health professionals and researchers through 16 regional education and research centers.

FDA approves prescription and over-the-counter drugs, including labeling requirements. The FDA Bureau of Radiological Health issues standards for radiation exposure and develops methods to control exposures. The FDA also regulates and classifies all medical devices under regulations published in 21 CFR 860 and 862 through 890. The Safe Medical Device Act of 1990 (SMDA) greatly expanded FDA authority to regulate medical devices. The FDA enforces other laws, including the Food, Drug, and Cosmetic Act, the Fair Packaging and Labeling Act, the portions of the Public Health Service Act relating to biological products (42 CFR, Parts 262–263), and the Radiation Control for Health and Safety Act, addressing lasers, x-rays, and microwaves (42 CFR 263).

The Agency for Healthcare Research and Quality (AHRQ) serves as the health services research arm of the DHHS. The agency provides a major source of funding and technical assistance for health services research and research training at leading educational institutions. Health services research examines how people get access to health care, how much care costs, and what happens to patients as a result of this care. The main goals of health services research are to identify the most effective ways to organize, manage, finance, and deliver high-quality care; reduce medical errors; and improve patient safety. The agency complements the biomedical research mission of the National Institutes of Health (NIH). The agency specializes in health-care-related research, including quality improvement and patient safety.

The Institute of Medicine (IOM) serves an independent scientific adviser. The agency provides advice based on evidence and grounded in science. The mission of the IOM embraces the health of people everywhere. As a part of the National Academies for Science, the agency focuses on matters such as biomedical science, medicine, and health. As a nonprofit organization, the institute works outside the framework of government to ensure scientifically informed analysis and independent guidance. The institute provides authoritative, evidence-based information to policy makers, professionals, and leaders in every sector of society. Every Institute of Medicine (IOM) report generated by committees goes through a review and evaluation process. The reviews conducted by expert panels ensure reviews remain unbiased. The institute focuses on reports or studies on subjects ranging from the quality of medical care to medical-error prevention. The institute also maintains a close working relationship with the NIH.

NIH operates 27 centers and serves as one of the world's foremost medical research organizations. NIH, an agency of DHHS, maintains facilities in Bethesda, Maryland. The agency serves as the focal point for federal health research. NIH pursues fundamental knowledge about the nature and behavior of living systems and the application of that knowledge to extend healthy lives by reducing the burdens of illness and disability. NIH provides leadership and direction to programs designed to improve health by conducting and supporting research related to the causes, diagnosis, prevention, and cure of human diseases. The institutes also seek to better understand the processes of human growth and development, the biological effects of environmental contaminants, and the understanding of mental, addictive, and physical disorders. The institutes also direct initiatives for collecting, disseminating, and exchanging information in the areas of health and medicine. The organization also helps with development of medical libraries and training of medical librarians and health information specialists.

The ATSDR works to protect adverse human health effects from hazardous substance exposures from waste sites, unplanned toxic releases, and pollution sources present in the environment. Congress intended ATSDR to perform functions that include public health assessments of waste sites, health consultations concerning specific hazardous substances, applied research in support of public-health assessments, information development/dissemination, and education/training concerning hazardous substances. The agency reports administratively to the CDC.

The Health Resources Services Administration (HRSA) seeks to improve and expand access to quality health care for all Americans. HRSA assures the availability of quality health care to low-income, uninsured, isolated, vulnerable, and special needs populations. HRSA seeks to eliminate barriers to care and health disparities by assuring quality of care, improving public health, and improving health-care systems. The agency manages guaranteed loan programs. It also provides architectural and engineering services during the application, design, bidding, and construction phases of federally assisted projects for the DHHS. HRSA oversees the federal "no-fault" system, which compensates individuals and families of those harmed by childhood vaccines, whether administered in the private or public sectors.

The Office of the Assistant Secretary for Preparedness and Response (ASPR), created by the Pandemic and All Hazards Preparedness Act, provides leadership in the areas of preventing, preparing, and responding to adverse health effects related to public health emergencies and disasters. ASPR focuses on preparedness planning and response, including coordinating the building of federal emergency medical operational capabilities. The office also works to promote countermeasures research and provide grants that will strengthen the capabilities of hospitals and health-care systems during emergencies and medical disasters. Under the Pandemic and All Hazards Preparedness Act, ASPR provides support to DHHS, which functions as the lead agency for the National Response Framework (NRF) for Emergency Support Function 8 (ESF).

The Centers for Medicare and Medicaid Services (CMS) oversees the Medicare and Medicaid programs and participation by health-care providers. The agency also manages surveys, certification, and quality-improvement responsibilities for health-care facilities. The agency publishes guidelines governing long-term nursing facilities and resident safety. The guidelines emphasize resident's rights and quality of care. CMS publishes enforceable survey standards. Refer to 42 CFR Part 483 to access these standards. The Omnibus Budget Reconciliation Act (OBRA) of 1987 gave the CMS the power to regulate all facilities receiving federal funds.

Department of Homeland Security

Created as a result of the September 11 terrorist attacks, the department consists of more than 20 previously disparate domestic agencies. These component agencies analyze threats and intelligence, guard our borders and airports, protect our critical infrastructure, and coordinate the response of our nation for future emergencies.

The Emergency Preparedness and Response Directorate provides oversight to domestic disaster-preparedness training and coordinate government disaster response. It coordinates the following agencies and resources: (1) Federal Emergency Management Agency (FEMA), (2) Strategic National Stockpile, (3) National Disaster Medical System, (4) Department of Energy (DOE) Nuclear Incident Response Team, (5) DOJ Domestic Emergency Support Teams, and (6) FBI National Domestic Preparedness Office.

The Science and Technology Directorate's mission focuses on using scientific and technological advances to help secure the homeland. The following assets play key roles in these efforts: (1) DOE Chemical, Biological, Radiological, and Nuclear (CBRN) Countermeasures Programs, (2) Environmental Measurements Laboratory of DOE, (3) National BW Defense Analysis Center of the Department of Defense, and (4) Plum Island Animal Disease Center, operated by the Department of Agriculture.

The Information Analysis and Infrastructure Protection Directorate helps analyze intelligence and information provided by other agencies involving threats to homeland security. The directorate also evaluates potential vulnerabilities of the nation's infrastructure by coordinating its efforts with (1) the Critical Infrastructure Assurance Office of the Department of Commerce, (2) the GSA Federal Computer Incident Response Center, (3) the DOD National Communications System, (4) the FBI National Infrastructure Protection Center, and (5) the DOE Energy Security and Assurance Program.

The Customs and Border Protection (CBP) seeks to prevent terrorists and terrorist weapons from entering the United States and ensuring the security of our nation at America's borders and ports of entry. CBP responsibilities include apprehending individuals attempting to enter the United States illegally. The agency seeks to stem the flow of illegal drugs and other contraband from entering the country. CBP also protects agricultural and economic interests from harmful pests and diseases. Other important functions involve regulating and facilitating international trade, collecting import duties, and enforcing U.S. trade laws.

The Federal Emergency Management Association (FEMA) moved to the Department of Homeland Security (DHS) in March 2003. FEMA goals include responding to, planning for, recovering from, and mitigating disasters. FEMA can trace its beginnings to the Congressional Act of 1803. FEMA maintains headquarters in Washington, DC, with regional and area offices across the country. FEMA operates the Mount Weather Emergency Operations Center and a training center in Emmetsburg, Maryland. FEMA works to build, sustain, and improve our capability to prepare for, protect against, respond to, recover from, and mitigate all hazards.

The US Fire Administration (USFA), a FEMA entity, provides leadership to foster a solid foundation for our fire and emergency services stakeholders in prevention, preparedness, and response. USFA focuses on four key areas, including the development of fire safety prevention education sessions in partnership with other federal agencies. It also promotes the professional development of those working in the fire and the emergency-response community. Finally, the agency assists state and local entities in collecting, analyzing, and disseminating data on the occurrence, control, and consequences of all types of fires.

The Immigration and Customs Enforcement (ICE) service functions as the principal investigative arm of DHS. The agency came into existence in 2003 through a merger of the investigative and interior enforcement elements of the US Customs Service and the Immigration and Naturalization Service (INS). ICE promotes homeland security and public safety through the criminal and civil enforcement of federal laws that govern border control, customs, trade, and immigration. The agency's Homeland Security Investigations (HSI) division and Enforcement and Removal Operations (ERO) divisions play key roles in ICE meeting its mission. ICE plans to take steps to streamline its management structure to give the agency a clearer sense of identity and focus.

The Transportation Security Administration (TSA) exists to protect transportation systems to ensure freedom of movement for people and commerce. TSA uses a layered security approach to help ensure the safety of the traveling public and transportation systems. Airport checkpoints constitute only one security layer of the many in place to protect aviation. Others include intelligence gathering and analysis, checking passenger manifests against watch lists, random canine team searches at airports, federal air marshals, federal flight-deck officers, and more security measures both visible and invisible to the public. TSA uses detection technology that advances security by the detection of dangerous materials. TSA works to harmonize screening measures and other security practices overseas in an effort to meet both international and US security standards. The agency's use of biometric identification allows verification of a person through fingerprints, iris scans, or a combination of the two.

The US Citizenship and Immigration Services (USCIS) provide oversight of lawful immigration into the United States. On March 1, 2003, USCIS officially assumed responsibility for the immigration service functions of the federal government. The Homeland Security Act of 2002 dismantled the former INS and separated the former agency into three components within DHS.

ICE and CBP handle immigration enforcement and border-security functions.

The US Coast Guard (USCG) operates as one of the five armed forces of the United States and the only military organization within DHS. The Coast Guard protects the maritime economy and the environment, defends our maritime borders, and saves those in peril. The service performs safety and environmental examinations of foreign vessels entering US ports. The service boards fishing vessels to perform safety and compliance checks. Working with the DOD, Homeland Security, and Department of Justice, the Coast Guard seizes illegal drugs bound toward the United States. The service also interdicts undocumented migrants attempting to illegally enter the United States. The service conducts security boarding of high-interest vessels bound for the United States. It also provides waterside security and escorts military freight conveyances supporting US military operations. The Coast Guard enforces laws that ensure safety on coastal, intracoastal, and inland navigable waterways. It operates the National Response Center that responds to reports of chemical or oil spills in navigable waterways. The service also oversees environmental cleanup activities on navigable waterways.

The US Secret Service was originally established in 1865, solely to suppress counterfeiting of US currency. In 1902, the Secret Service assumed full-time responsibility for protection of the president. The agency's primary investigative mission relates to safeguarding the payment and financial systems of the US. This mission historically relied on the enforcement of counterfeiting statutes. The Secret Service's investigative responsibilities now include crimes involving financial institution fraud, computer and telecommunications fraud, false identification documents, access device fraud, advance fee fraud, and electronic funds transfers and money laundering as it relates to the agency's core violations. To combat these crimes, the Secret Service uses a proactive approach that utilizes advanced technologies. The Patriot Act increased the Secret Service's role in investigating fraud and related activity related to the use of computers. The service transferred from the Department of the Treasury to DHS on March 1, 2003.

National Transportation Safety Board

The National Transportation Safety Board (NTSB), an independent agency, investigates every civil aviation accident in the United States and significant accidents in other modes of transportation, including highway, marine, pipeline, and railroad. The agency's authority comes from Title 49 of the United States Code, Chapter 11. The Board of the NTSB consists of five presidentially appointed members, each confirmed by the Senate for a 5-year term. The NTSB also conducts safety studies and evaluates the effectiveness of other government agencies' programs for preventing transportation accidents. It reviews the appeals of enforcement actions involving aviation and mariner certificates issued by the FAA and the USCG.

Chemical Safety Board

The U.S. Chemical Safety Board (CSB) became operational in January 1998. The CSB operates as an independent agency charged with investigating industrial chemical accidents. The president appoints board members, with confirmation required by the Senate. CSB conducts root-cause analyses and investigations of chemical accidents at fixed industrial facilities. The agency does not issue fines or citations but does make recommendations. Congress designed the CSB to be nonregulatory and independent of other agencies. This permits investigations to also determine the effectiveness of compliance regulations and regulatory enforcement. The CSB investigative staff includes chemical and mechanical engineers, industrial safety experts, and other specialists.

Consumer Product Safety Commission

The Consumer Product Safety Act of 1972 established the commission to protect the public against unreasonable risks of injuries associated with consumer products. The Consumer Product Safety Commission (CPSC) operates as an independent agency headed by five commissioners nominated by the president and confirmed by the Senate. CPSC monitors the safety of over 15,000 kinds of consumer products. The CPSC works to ensure the safety of consumer products such as toys, cribs, power tools, cigarette lighters, and household chemicals. CPSC announces recalls of products that present a significant risk to consumers either because the product may be defective or because it violates a mandatory standard issued by CPSC. The agency does not possess any jurisdiction over automobiles or other on-road vehicles, tires, boats, alcohol, tobacco, firearms, food, drugs, cosmetics, pesticides, and medical devices. The Federal Trade Commission handles complaints of false advertising, fraud, and product quality.

Consensus Standards Organizations

American Conference of Governmental Industrial Hygienists

The National Conference of Governmental Industrial Hygienists (NCGIH) first convened in June 1938 and in 1946 changed its name to the American Conference of Governmental Industrial Hygienists (ACGIH). ACGIH uses nine committees to focus on a range of topics in agricultural safety and health, air sampling instruments, bioaerosols, biological exposure indices (BEI), industrial ventilation, and international and small business. ACGIH publishes Threshold Limit Values (TLVs) for chemical substances (TLVs®-CS), and TLVs for physical agents (TLVs®-PA). The list of TLVs® includes more than 640 chemical substances and physical agents plus more than 40 BEIs®.

American National Standards Institute

The organization was founded in 1918 for the purpose of consolidating voluntary standards. American National Standards Institute (ANSI) functions as a federation of more than 1500 professional, trade, governmental, industrial, labor, and consumer organizations. It publishes consensus standards developed by various technical, professional, trade, and consumer organizations. ANSI also serves as the coordinating agency for safety standards that have been adopted for international implementation. OSHA adopts many ANSI standards, and industries can adopt standards related to their operations. ANSI provides members access to more than 9000 standards from around the world. ANSI publishes specifications for PPE, including safety glasses, goggles, hard hats, safety shoes, fall-protection equipment, eyewash stations, and emergency eye-wash and shower equipment.

American Society of Heating, Refrigerating, and Air-Conditioning Engineers

American Society of Heating, Refrigerating, and Air-Conditioning Engineers (ASHRAE) fulfills its mission of advancing heating, ventilation, air conditioning, and refrigeration to serve humanity and promote a sustainable world through research, standards writing, publishing, and continuing education. ASHRAE serves as the foremost source of technical and educational information and the primary provider of opportunity for professional growth in the arts and sciences of heating, ventilation, air conditioning, and refrigerating. Through its membership, ASHRAE writes standards that set uniform methods of testing and rating of equipment. It also establishes accepted practices for the HVAC&R industry worldwide, including publishing standards on the design of energy-efficient buildings and ventilation.

American Society of Testing Materials International

American Society of Testing Materials (ASTM) is the world's largest source of voluntary consensus standards and publishes more than 8000 standards annually. Standards address broad categories of topics such as medical devices, occupational safety and health, environmental effects, energy, and security systems. The standards address six major categories of information: (1) classification information on materials grouped together by characteristics, (2) practices and procedures detailing how to accomplish a process or function, (3) testing methods for specific products or materials, (4) guides to provide directional procedures, (5) precise information about material specifications, and (6) terminology or a compilation of definitions and terms.

Compressed Gas Association

Compressed Gas Association (CGA) develops technical and safety standards for the compressed gas industry. Members work together through a committee system to develop technical specifications, safety standards, and educational materials, and to promote compliance with regulations and standards in the workplace. Member companies represent manufacturers, distributors, suppliers, and transporters of gases, cryogenic liquids, and related products. The CGA publishes more than 100 technical standards, many of which are formally recognized by US government agencies. The association publishes the *Handbook of Compressed Gases*, which is widely used and sets forth the recognized safe methods for handling, storing, and transporting industrial gases.

Factory Mutual Research Corporation

Factory Mutual Research Corporation (FM), a nationally recognized testing laboratory and approval organization recognized by OSHA, focuses on industrial loss-control issues. FM also performs third-party testing on fire-extinguishing equipment, sprinklers, building materials, and smoke detectors. FM lists approved equipment, materials, and other services in its annual 500-page guide. Manufacturers can display a special symbol on approved items to inform users and buyers that the product or piece of equipment has been tested and approved by an independent laboratory. OSHA recognizes the FM lab.

International Electrotechnical Commission

This commission serves as a leading global organization that prepares and publishes international standards for all electrical, electronic, and related technologies. The International Electrotechnical Commission (IEC) charter embraces the areas of electronic, magnetic and electromagnetic, electro-acoustic, multimedia, telecommunication, and energy production and distribution technologies. It also supports associated general disciplines such as terminology and symbols, electromagnetic compatibility, measurement, performance, dependability, design, development, safety, and the environment. IEC standards provide industry and users with the framework for economies of design, greater product and service quality, more interoperability, and better production and delivery efficiency. Using IEC standards for certification at a national level ensures that a certified product meets good manufacturing standards.

National Council on Radiation Protection and Measurements

The National Council on Radiation Protection and Measurements (NCRP) performs functions related to radiation protection and measurements. Founded as the Advisory Committee on X-Ray and Radium Protection in 1929, it originally represented all national radiological organizations in the United States. The NCRP originally operated in an informal manner when making disseminating information and recommendations on radiation protection. The NCRP reorganized and received a charter from the US Congress in 1964.

National Fire Protection Association

National Fire Protection Association (NFPA) serves as the world's leading advocate of fire prevention. The association publishes about 300 safety codes and standards. NFPA encourages the broadest possible participation in its consensus code-development process. NFPA relies on more than 6000 volunteers from diverse professional backgrounds to serve on over 200 technical code and standard development committees. The code development process holds ANSI accreditation. Some examples of NFPA codes relevant to health care include NFPA 70, National Electrical Code®, NFPA 99 Healthcare Facilities, and NFPA 101, Life Safety Code®. NFPA offers excellent education and training programs dealing with the latest fire and life safety requirements, technologies, and practices. NFPA also administers several professional certification programs, including Certified Fire Protection Specialist, Certified Fire Inspector, and Certified Fire Plans Examiner. NFPA develops dozens of texts, guides, and other materials to assist firefighters and first responders.

National Sanitation Foundation International

Founded in 1944, the National Sanitation Foundation (NSF) develops standards, conducts product testing, and provides certification services in areas related to public health and safety. Technical resources include the operation of physical and performance testing facilities and analytical chemistry and microbiology laboratories. NSF International operates as a not-for-profit nongovernmental organization. NSF develops national standards, provides learning opportunities, and provides third-party conformity assessment services. The foundation known and recognized for its scientific and technical expertise continues to focus on a variety of pertinent issues related to health and environmental sciences.

Safety Equipment Institute

Safety Equipment Institute (SEI), a private nonprofit organization established in 1981, administers nongovernmental third-party certification programs to test and certify a broad range of safety equipment products. ANSI accredits SEI certification programs, which function as voluntary and are available to any manufacturer of safety and protective equipment seeking product certification. SEI focuses efforts on continuous product testing and quality-assurance audits that result in products earning certification. Product testing meets requirements established by voluntary, governmental, and other consensus standards relevant for a given product. A number of organizations such as ANSI, ASTM, and NFPA promulgate standards based on SEI certification.

Underwriters Laboratories

Underwriters Laboratories (UL), a nonprofit organization, maintains laboratories for the examination and testing of systems, devices, and materials to ensure compliance with safety and health standards. UL inspects or tests more than 70,000 products each year, including fire-fighting equipment, lockout/tagout supplies, lighting fixtures, and flammable liquid storage containers. UL certification pertains only to the area of safety and does not focus on overall product performance testing. UL has issued more than 500 standards, with many adopted by ANSI. UL publishes directories of companies whose products meet or exceed criteria outlined in appropriate standards.

National Board of Boiler and Pressure Vessel Inspectors

This board promotes safety and uniformity in the construction, installation, repair, maintenance, and inspection of pressure equipment. It promotes safety and educates the public and government officials on the need for manufacturing, maintenance, and repair standards. The board offers comprehensive training and education for inspectors and pressure equipment maintenance professionals. The board promotes an inspection process through the commissioning of inspectors. Inspectors must pass a comprehensive examination administered by the national board. The board also sets worldwide industry standards for pressure relief devices and other appurtenances. The board investigates pressure-equipment incidents involving code compliance. The board develops installation, inspection, repair, and alteration standards for the National Inspection Code.

Professional Society and Membership Associations

Society for Human Resource Management

The Society for Human Resource Management (SHRM), founded in 1948, provides services to more than 60,000 members. SHRM produces several professional publications, including the monthly *HR Magazine*. The society promotes a certification program administered by the Human Resource Certification Institute. Candidates may apply for certification as a professional in human resources (SPHR) or senior professional in human resources (SPHR). Certification requires experience in an exempt HR position and successful completion of a certification examination. The society holds an annual conference and exposition, an employment law and legislative conference, a leadership conference, and a diversity conference.

American Association of Occupational Health Nurses

The American Association of Occupational Health Nurses (AAOHN) serves the largest group of health-care professionals in the workplace. The association seeks to advance and maximize the health, safety, and productivity of domestic and global workforces. The association provides education, research, public policy, and practice resources for occupational and environmental health nurses. AAOHN advances the profession of occupational and environmental health nursing by focusing on education and research, professional practice/ethics, communications, governmental issues, and creating alliances.

Association of Occupational Health Professionals

This association works to define employee health issues and serve as a leading advocate for occupational health professionals serving in health-care organizations. The association participates in governmental affairs and meets with OSHA, NIOSH, and congressional representatives to address association positions. The association sponsors an annual national conference where members meet to share, network, and attend professional education sessions. OSHA and the association recently entered into an alliance to promote worker health and safety in health care.

American Chemistry Council

The American Chemistry Council represents companies that make the products to improve life while protecting the environment, public health, and the security of the country. Founded in 1872, the member companies must commit to implement a set of goals and guidelines that exceed federal safety, security, and environmental regulations. The American Chemistry Council seeks to provide exceptional advocacy based on improved member performance, high-quality scientific research, communications, and effective participation in political processes.

American Health Care Association

The American Health Care Association, a nonprofit federation of affiliated state health organizations, represents nearly 12,000 assisted-living, nursing, and subacute care providers caring for more than 1.5 million individuals. The association represents the long-term care community to government, business leaders, and the public. The association maintains legislative, regulatory, public affairs, and member services staffs to serve the interests of government, the public, and member providers. The association seeks to focus on providing quality care to the nation's frail, elderly, and disabled.

American Hospital Association

The American Hospital Association (AHA) represents and serves all types of hospitals, health-care networks, their patients, and local communities. AHA uses representation and advocacy activities to ensure members' perspectives and needs receive fair treatment in national health policy development, legislative and regulatory debates, and judicial matters. Founded in 1898, the AHA provides education for health-care leaders and serves as an informational resource on health-care issues and trends.

American Industrial Hygiene Association

The American Industrial Hygiene Association® (AIHA) serves the needs of occupational and environmental health and safety professionals practicing industrial hygiene. Founded in 1939, AIHA® is a nonprofit organization with 73 local sections. Members practice in industry, government, labor, academic institutions, and independent organizations. The association works with the American Board of Industrial Hygiene to promote certification of industrial hygienists. The association also provides educational opportunities that help keep occupational and environmental health and safety professionals current in the field of industrial hygiene. The association operates several highly recognized laboratory accreditation programs.

American Osteopathic Association

The association represents more than 47,000 osteopathic physicians. American Osteopathic Association (AOA) also promotes public health, encourages scientific research, and serves as the primary certifying body for osteopath doctors. The association serves as the accrediting agency for all osteopathic medical schools and health-care facilities, including acute-care hospitals. The federal government granted AOA "deeming authority" to accredit laboratories within hospitals under the Clinical Laboratory Improvement Amendments of 1988. The association offers accreditation for ambulatory care, surgery, mental health, substance abuse, and physical rehabilitation medicine facilities.

American Public Health Association

The American Public Health Association (APHA), established in 1872, functions as the oldest and most diverse organization of public health professionals in the world. The association seeks to help protect all American families and their communities from preventable serious health threats. The association strives to ensure the accessibility of community-based health promotions, disease-prevention activities, and preventive health services for all Americans. The association promotes the scientific and professional foundations related to public health practice and policy. APHA also advocates the importance of a healthy global society by emphasizing prevention and enhancing members' ability to promote environmental and community health.

American Society of Healthcare Engineering

As an AHA-affiliated organization with members worldwide, the American Society of Healthcare Engineering (ASHE) promotes the importance of maintaining effective safety, emergency preparedness, engineering, and security functions within health-care organizations. ASHE takes a leading role in providing members with information about regulatory codes and standards. ASHE promotes health-care education through professional development seminars and conferences. Monthly publication of technical documents keeps members informed on the latest changes and developments related to health-care engineering and facility management.

American Society of Industrial Security

The American Society of Industrial Security (ASIS), the largest organization for security professionals, works to increase the effectiveness and productivity of security personnel serving in a variety of roles worldwide. ASIS advocates the important role that effective security measures can play in the success of any of organization, including retail businesses, industrial complexes, cyber-oriented endeavors, and governmental entities. ASIS provides members and the security community with access to a full range of services, including the industry's top-rated magazine, *Security Management*. ASIS works to lead the way for advanced and improved security performance.

American Society of Safety Engineers

The society was founded in October 1911 as the United Society of Casualty Inspectors, and functions as the largest member organization of individual safety professionals. It works to promote the safety profession and foster the professional development of its members. The American Society of Safety Engineers (ASSE) plays an important role in the development of many national programs and standards. The objectives of the society's 138 chapters focus on promoting, establishing, and maintaining standards for the safety profession. The society also publishes the monthly peer-reviewed journal *Safety Professional*. ASSE sponsors a number of educational conferences, seminars, and educational programs.

American Welding Society

The American Welding Society (AWS), founded in 1919, seeks to promote the advancement of the science and technology of welding and related disciplines. From factory floor to high-rise construction, from military weaponry to home products, AWS continues to lead the way in supporting welding education and technology development to ensure a strong and competitive America.

CHEMTREC

CHEMTREC, established in 1971, provides a public service hotline for those needing information and assistance with incidents involving chemicals or hazardous materials releases. Registration with CHEMTREC authorizes shippers of hazardous materials the right to portray the CHEMTREC phone numbers on their shipping documents, Safety Data Sheets, and hazard communications labels. The portrayal of the CHEMTREC phone numbers helps registrants to comply with DOT requirements found in 49 CFR 172.604. DOT requires shippers of hazardous materials to provide a 24-hour emergency telephone number on shipping documents. CHEMTREC operates around-the-clock.

ECRI Institute

ECRI operates as a nonprofit institute with worldwide recognition as a leading independent organization totally committed to advancing the quality of health care. ECRI provides information services and technical assistance to more than 5000 health care-related organizations. ECRI maintains over 30 databases, publications, information systems, and technical-assistance services. ECRI provides alerts related to technology hazards and the results of medical product or technology assessments.

Ergonomics Society

The Ergonomics Society, a UK-based professional society, serves ergonomists, human factors professionals, and others involved in user-centered design functions. The society promotes using knowledge about human abilities and limitations to promote improved designs for comfort, efficiency, productivity, and safety. The society serves the interests of members by encouraging and maintaining high standards of professional practice through education, accreditation, and professional development. The society also promotes the interests of ergonomics and ergonomists serving in government, academia, business, and industry.

Health Physics Society

The society, established in 1956, serves as a scientific organization of professionals who specialize in radiation safety. The society's members represent all scientific and technical areas related to radiation safety, including academia, government, medicine, research and development, analytical services, consulting, and industry. The society, as an independent, nonprofit scientific organization, does not maintain any affiliation with any government or industrial organization. The society promotes public information preparation and dissemination, education and training opportunities, and scientific information exchange through conferences and meetings.

Human Factors and Ergonomics Society

This society advocates the use of human factor knowledge to achieve compatibility in the design of interactive systems consisting of people, machines, and operational environments. The society encourages appropriate education and training for those entering the human factors profession. The bimonthly journal *Human Factors* presents original papers of scientific merit that contribute to understanding and application of human factors. The journal features articles on methodology, procedures, literature reviews, technical research results, and papers of general professional interest.

International Association of Nanotechnology

The International Association of Nanotechnology (IANT) focuses on fostering scientific research and business development in the areas of nanoscience and nanotechnology. IANT does not endorse or support any applications that use and misuse advanced technology for destructive purposes. IANT promotes developing a roadmap and framework for the use of the emerging technology, including various issues relating to nanotechnology.

International Consumer Product Health and Safety Organization

The International Consumer Product Health and Safety Organization (ICPHSO), founded in 1993 as a not-for-profit organization, works to address the many health and safety issues related to consumer products marketed globally. ICPHSO attracts international members, who meet annually to exchange ideas and share information. ICPHSO members, represent government agencies, manufacturers, importers, retailers, certification organizations, testing laboratories, law firms, academia, standards organizations, media, and consumer advocacy groups.

International Organization for Standardization

The International Organization for Standardization (ISO) publishes thousands of international standards that address topics such as agriculture, construction, mechanical engineering, medical devices, and the newest information technology developments. ISO functions using a network of the national standards institutes located in 163 countries. A central secretariat located in Geneva, Switzerland, coordinates the entire system. ISO forms a bridge between public and private interests. ISO promotes consensus on solutions that meet both the requirements of business and the broader needs of society. ISO objectives focus on ensuring that products, systems, machinery, and devices work well and safely.

International System Safety Society

The International System Safety Society promotes the safe development of complex systems, products, and processes. The society stresses the use of safety principles during every phase of a design project. A complex system requires the use or application of system engineering principles, management concepts, hazard controls, and risk analyses during all development phases. The practice of system safety includes professionals with a range of specialties, including engineering, human factors, science, safety science, law, and management. The society draws members from around the world and maintains affiliations with major corporations, educational institutions, and other organizations with similar objectives.

National Restaurant Association

The National Restaurant Association (NRA) serves more than 380,000 businesses, including restaurants, suppliers, educators, and nonprofit organizations. The members of the NRA come from every corner of the restaurant and hospitality industry. The NRA works to improve food safety and security within the restaurant industry. NRA also strives to promote increasing restaurant nutrition information availability and consumer awareness. The association prepares and distributes educational materials, including self-inspection guidelines, safety concerns, OSHA requirements, and fire protection.

National Safety Council

The council, a not-for-profit organization, promotes safety throughout the United States. The National Safety Council (NSC) recently announced its new SMS. The system can provide companies with a structure to implement a comprehensive and balanced safety management function. Workplace safety and transportation safety remain critical areas of focus. The NSC saves lives by preventing injuries and deaths at work, in homes and communities, and on the roads through the use of leadership, research, education, and advocacy. The council vision statement simply says, Making Our World Safer.

Risk and Insurance Management Society

The Risk and Insurance Management Society (RIMS), a not-for-profit organization, advances the practice of risk management, a professional discipline to protect physical, financial, and human resources. Founded in 1950, RIMS represents nearly 4000 industrial, service, nonprofit, charitable, and governmental entities. The society serves more than 10,000 risk-management professionals around the world. RIMS serves its members by proactively providing the highest quality products, services, and information to manage risk effectively.

Review Exercises

1. Why should hazard control managers have a good working knowledge and understanding of government agencies, consensus standard organizations, and professional associations?
2. What is the purpose of the FR, and how does it work?
3. Explain the role the CFR plays in governmental compliance.
4. What two agencies were created by the OSH Act?
5. List the five areas or industry segments addressed by published OSHA safety and health standards.
6. List OSHA's top three inspection priorities.
7. Describe the purpose and basic operation of OSHRC.
8. What primary law addressing disposal activities is enforced by EPA standards?
9. Describe the scope of the Toxic Substance Control Act.
10. Explain the mission of the Federal Motor Carrier Safety Administration.
11. Which agency maintains the NDRF as a ready source of ships for use during a national emergency?
12. What are the "4Es" of safety as listed by the FHWA?

13. What are four of the six elements considered by MSHA when issuing citations?
14. List the five types of facilities regulated by the NRC.
15. List five government agencies operated by the Department of Health and Human Services.
16. List the five organizations operated by the DHS's Emergency Preparedness and Response Directorate.
17. Explain the mission and operation of the Transportation Safety Board.
18. Which organization issues TLVs?
19. Describe the purpose or mission of the following consensus standard organizations:
    a. ANSI
    b. ASHRAE
    c. ASTM
    d. FM
    e. IEC
    f. NFPA
    g. NSF
    h. SEI
    i. UL

Chapter 6

Managing Hazardous Materials

Introduction

This chapter provides an overview of hazardous materials and substances. The Institute of Hazardous Materials Management (IHMM) defines a hazardous material "as any item or agent to include biological, chemical, or physical with potential to cause harm to humans, animals, or the environment." This harm can occur from the material itself or through interaction with other factors. Several federal agencies define and regulate hazardous materials and substances. The primary agencies include the Environmental Protection Agency (EPA), the Occupational Safety and Health Administration (OSHA), the Department of Transportation (DOT), and the Nuclear Regulatory Commission (NRC).

Hazardous substances can enter the body through the skin, respiratory system, the mouth, and the eyes. Some substances can also damage the skin or eyes directly without being absorbed. Inorganic lead can be inhaled or swallowed, but it does not penetrate the skin. Sometimes a chemical substance can enter through more than one route. Exposures to hazardous materials cause stress on the body if inhaled, absorbed, or ingested. Exposure effects depend on concentration, duration of exposure and route of exposure, physical properties, and chemical properties. Other chemicals and physical agents can influence the effects exerted by a hazardous substance. Organizations should train and educate individuals in proper ways to handle, store, use, and segregate hazardous materials and wastes. When evaluating hazardous risks, consider the concentration of substances used, duration of exposures, ventilation effectiveness, reactivity potential, flammability, and human health hazards. Exposure sources and other harmful factors relate to the type of processes, technologies, products, and equipment used in the workplace. Consider ways to eliminate or mitigate potential exposures, including the substitution of a less-hazardous substance. Consider the use of proven engineering controls to protect and separate persons from harmful exposures.

Definitions

OSHA defines as hazardous any substance or chemical with a "health hazard" or "physical hazard," including substances that are carcinogens, toxic agents, irritants, corrosive, or sensitizers. EPA incorporates the OSHA definition and adds any item or chemical that can cause harm to people, plants, or animals when released by spilling, leaking, pumping, pouring, emitting, emptying, discharging, injecting, escaping, leaching, dumping, or disposing into the environment. DOT defines a hazardous material as any item or chemical being transported or moved that poses a risk to public safety or the environment. The NRC regulates items or chemicals that are "special nuclear source" or by-product materials or radioactive substances.

Organizations must identify hazardous materials requiring special handling procedures. Organizations must also minimize the risks during use and ensure the use of proper disposal methods. Organizations must implement written programs to ensure the management and disposal of all hazardous materials. OSHA requires development of a written hazard communication plan to address hazardous materials used in the workplace.

The EPA defines four types of hazardous materials in the Resource Conservation and Recovery Act (RCRA). *Corrosive materials* possess the ability to degrade the structure or integrity of substances, objects, or materials. *Ignitable materials* can readily burn or ignite, including some materials with the potential to auto-ignite upon contact with the air. A *reactive material* possesses the ability to readily combine with other substances to produce a sudden or violent release of heat/energy. RCRA defines a *toxic material* as any substance capable of damaging the environment or of causing illness or death in humans, animals, fish, and plants.

OSHA published the Air Contaminant Standards in 29 CFR 1910, Subpart Z. A permissible exposure limit (PEL) establishes the maximum allowable OSHA exposure. The limit considers the total exposure received during a monitoring period of 8 hours for individuals that work 40 hours a week. OSHA defines a short-term exposure limit (STEL) as an "excursion" exposure allowed at one time (normally measured in a 15-minute period). The National Institute for Occupational Safety and Health (NIOSH) publishes recommended exposure limits (RELs) for selected hazardous substances. Refer to the NIOSH website for additional information. A ceiling limit refers to the amount of airborne exposure that should never be exceeded. The American Conference of Governmental Industrial Hygienists (ACGIH) publishes voluntary threshold limit values (TLVs) for use by industrial hygienists. Base the toxicity of any exposure on its duration, level of the hazard, and individual susceptibility. Acute exposures occur from short-term exposure of high concentrations, resulting in irritation, illness, or death. Chronic effects involve continued exposure to a toxic substance over a long period of time. Refer to the OSHA Additive Formula found in 29 CFR 1910.1000 for use when determining exposure effects of a substance containing two or more hazardous ingredients.

Respiratory Protection (29 CFR 1910.134)

Respirators prevent the inhalation of harmful airborne substances and provide fresh air in oxygen-deficient environments. An effective respiratory protection plan must address the following: (1) hazards encountered, (2) type and degree of protection needed, (3) medical evaluation for respirator usage, (4) selection and fit requirements, (5) training on use and care, and (6) methods to ensure continued program effectiveness.

Types of Respirators

Air-purifying respirators come in either full-face or half-mask versions. These types of respirators use a mechanical or chemical cartridge to filter dust, mists, fumes, vapors, or gaseous substances. Only use disposable air-purifying respirators once or until the cartridge expires. These respirators contain permanent cartridges with no replaceable parts. Reusable air-purifying respirators use both replaceable cartridges and parts. The replaceable cartridges and parts must come from the same manufacturer to retain a NIOSH approval.

Disposable or reusable air-purifying respirators contain no replaceable parts except cartridges. Gas masks, designed for slightly higher concentrations of organic vapors, gases, dusts, mists, or fumes, use a volume of sorbent much higher than a chemical cartridge. Powered air-purifying respirators use a blower to pass the contaminated air through a filter. The purified air is then delivered into a mask or hood. These respirators filter dusts, mists, fumes vapors, or gases like other air-purifying respirators. Never use air-purifying respirators in any oxygen-deficient atmospheres. Oxygen levels below 19.5% require either a source of supplied air or a supplied-air respirator. Levels below 16% are considered to be unsafe and could cause death. Supplied-air respirators provide the highest level of protection against highly toxic and unknown materials. Supplied air refers to self-contained breathing apparatuses and airline respirators. Airline respirators have an air hose connected to a fresh air supply from a central source. The source comes from a compressed air cylinder or air compressor that provides breathable air. Emergency Escape Breathing Apparatuses provide oxygen for short periods of times such as 5 or 10 minutes, depending on the unit. Permit these devices only for emergency situations such as escaping from environments with immediately dangerous to life or health (IDLH) conditions.

Determining Cartridge Type and Selecting Respirators

Determine the correct cartridge for air-purifying respirators by contacting a respirator professional or referring to the Safety Data Sheet (SDS) of the substance needing filtering. Cartridges use a color scheme designating the contaminant needing filtering. Replace the cartridge any time a wearer detects odor, irritation, or taste of a contaminant. The proper selection and use of a respirator depends on an initial determination of the concentration of the hazard or hazards present in the workplace or area with an oxygen-deficient atmosphere.

IDLH atmospheres pose the most danger to workers. Use a full-facepiece, pressure-demand self-contained breathing apparatus (SCBA) or a combination full-facepiece, pressure-demand supplied-air respirator (SAR) for dangerous atmospheres. Respirator selection requires matching the respirator with the degree of hazard and needs of the user. Choose only devices that fully protect the worker and permit job accomplishment with minimum physical stress (Table 6.1).

Medical Evaluations

Persons assigned to tasks requiring the use of a respirator must possess the physical ability to work while using the device. OSHA requires employers to ensure the medical fitness of individuals who must wear respirators. The fitness evaluation considers the physical and psychological stress imposed by a respirator. It must also evaluate the stress originating from job performance. Employers must ensure that employees pass the evaluation before fit testing or permitting use of the respirator for the first time. A physician or other licensed health-care professional must determine medical eligibility for respirator wear. A qualified health-care provider includes physicians, occupational health nurses, nurse practitioners, and physician assistants, if licensed to do so in the state in which they practice.

Table 6.1   Respirator Selection Considerations

- Physical and chemical properties of the air contaminant
- Concentration of the contaminant
- Permissible exposure limits
- Nature of the work operation or process
- Length of time respirator worn
- Work activities and physical/psychological stress
- Fit testing, functional capabilities, and limitations of the respirator
- Characteristic or form of gas, dust, organic vapor, fume, mist, oxygen deficiency, or any combination

Fit Testing

The OSHA standard requires the fit testing of all tight-fitting respirators. OSHA does not exclude disposable particulate respirators from fit-testing requirements. Some employees may not achieve an adequate fit with certain respirator models or a particular type of respirator. Provide alternative respirator choices to ensure worker protection. Employers must provide a sufficient number of respirator models and sizes from which employees can choose an acceptable respirator with a correct fit. Quantitative fit test (QNFT) refers to the assessment of the adequacy of respirator fit by numerically measuring the amount of leakage into the respirator. A QNFT uses an instrument to take samples from the wearer's breathing zone. Adhere to the OSHA protocol for a QNFT as detailed in Appendix A of 29 CFR 1910.134. A qualitative fit test (QLFT) refers to a pass/fail test that assesses adequacy of respirator fit that relies on the individual's response to a test agent. A QLFT, according to 29 CFR 1910.134, applies only to negative-pressure air-purifying respirators that must achieve a fit factor of 100 or less. Since the QLFT relies on the subjective response of the wearer, accuracy may vary.

Training and Information

The proper fit, usage, and maintenance of respirators remain the key elements that help ensure employee protection. Train employees about the proper use of respirators and the general requirements of the Respiratory Protection Standard. Training must address employer obligations such as written plans, respirator selection procedures, respirator use evaluation, and medical evaluations. Employers must ensure proper maintenance, storage, and cleaning of all respirators. They must also retain and provide access to specific records as required by OSHA. Employees must know basic employer obligations as related to their protection. OSHA requires annual training of all workers expected to wear a respirator. New employees must attend respirator training before using a respirator in the workplace.

Program Evaluation

Employers must conduct workplace evaluations to ensure the scope of the written respirator plan protects those required to use respirators. Employers must also evaluate the continued effectiveness of the written respirator plan. Proper evaluations help to determine if workers use and wear respirators correctly. The evaluations can also indicate the effectiveness of respirator training. Employers must solicit employee views about respirator plan effectiveness and determine any problem areas.

Recordkeeping Requirements

OSHA requires employers retain written information regarding medical evaluations, fit testing, and the respirator plan effectiveness. Maintaining this information promotes greater employee involvement and provides compliance documentation. Employers must retain a record for each employee subject to medical evaluation. This record includes results of the medical questionnaire and, if applicable, a copy of the health-care professional's written opinion. Maintain records related to recommendations, including the results of relevant examinations and tests. Records of medical evaluations must be retained as required by 29 CFR 1910.1020, Access to Employee Exposure and Medical Records. Retain fit-test records on users until administration of the next test.

Physical Properties of Hazardous Materials

The physical properties of a chemical substance can include characteristics such as vapor pressure, solubility in water, boiling point, melting point, molecular weight, specific gravity, and flash point. Vapor refers to the gaseous state of a substance. Vapors can combine with the oxygen to form a mixture that can ignite or explode. Vapor density refers to the ratio of the weight of a volume of vapor or gas to the weight of an equal volume of clean but dry air. Vapor densities less than 1.0 tend to rise and spread out. Vapor densities greater than 1.0, or heavier than air, tend to sink to the point near the ground. These vapors can then travel along the ground—sometimes for long distances—and find ignition sources. This makes chemicals with high vapor densities particularly dangerous.

We can define the flash point of a liquid as the lowest temperature at which it gives off enough vapors to form an ignitable mixture with surrounding air. An ignition source is anything that can cause something to burn. Common ignition sources include sparks from tools and equipment, open flames, hot metal from grinding or welding operations, electric coils, and overheated bearings. The flowing of flammable liquids can result in the buildup of static electricity. Proper grounding ensures that an electrical charge goes to the ground rather than building up on the drum of flammable or combustible material. Bonding refers to a process that equalizes the electrical charge between the drum and the transfer container, reducing buildup of electrical charges on one of the containers. We can define ignition temperature as the minimum temperature at which a chemical will burn and continue burning without an ignition source. The rate of combustion remains the key difference between flammable and explosive substances. *Fire* refers to a process with a rapid release of energy and a high combustion rate. *Explosion* refers to the instantaneous release of energy and involves an extremely rapid rate of combustion.

Airborne Exposure

An exposure of an individual relates directly to the concentration of a hazardous substance as related to the per unit volume of air. We usually express airborne concentrations in terms of milligrams of substance per cubic meter of air (mg/m^3) or parts of substance per million parts (ppm) of air. Express asbestos and other airborne fibers by using per cubic centimeter (f/cc) or fibers per cubic meter (f/m³) of air. OSHA requires consideration of feasible administrative or engineering controls to reduce exposure risks. When these controls prove ineffective, organizations must use personal protective equipment (PPE) or other protective measures to protect employees. Ensure the use of any equipment and/or technical measures receives approval from a competent industrial hygienist or other technically qualified person. Subpart Z contains exposure limit Tables Z-1, Z-2, or Z-3 for substances not covered by a specific standard.

OSHA Additive Formula

OSHA provides an "additive formula" in 29 CFR 1910.1000 for computing exposure to a substance containing two or more hazardous ingredients. Employers must monitor and compute the equivalent exposure using the following formula:

- $E(m)$ is the equivalent exposure for the mixture
- C is the concentration of a particular contaminant
- L is the exposure limit for that substance specified in Subpart Z
- Value of $E(m)$ shall not exceed unity

To illustrate the formula prescribed in paragraph (d) (2) (I) of Subpart Z, consider the following exposures:

Substance A—Actual exposure at 500 ppm with a PEL of 1000 ppm
Substance C—Actual exposure at 45 ppm with a PEL of 200 ppm
Substance C—Actual exposure at 40 ppm with a PEL of 200 ppm

Substituting the above values into the formula gives the following results:

$E(m) = 500/1000 + 45/200 + 40/200$
$E(m) = 0.500 + 0.225 + 0.200$
$E(m) = 0.925$

Since $E(m)$ is less than unity, the exposure combination is within acceptable limits. If the value exceeds 1, the exposure would have been above the acceptable limit.

Emergency Showers and Eyewashes

OSHA standard (29 CFR 1910.151) requires employers to provide suitable facilities for quick drenching of the eyes and body for individuals exposed to corrosive materials. OSHA does not specify minimum operating requirements or installation setup requirements. American National Standards Institute (ANSI) Standard Z358.1 underwent revision in 2009, led by the efforts of the International Safety Equipment Association (ISEA). Approved by ANSI, the standard became known as ANSI/ISEA Z358.1-2009. Organizations should ensure flushing fluids remain clear and free from foreign particles.

For self-contained units, manufacturers provide suggested fluid-replacement guidelines. Preservatives can help control bacteria levels in flushing fluids. A preservative's performance depends on several factors, including the initial bacterial load of the water and a potential biofilm

Table 6.2   Basic Requirements for Eyewash and Shower Facilities

- Valves must activate in 1 second or less.
- Installed 10 seconds from the hazard.
- Located in a lighted area and identified with a sign.
- Train workers on equipment use and appropriate PPE.
- Activate plumbed units weekly.
- Maintain self-contained units according to manufacturers' specifications.

in the station. Self-contained eyewash stations should be drained completely, disinfected, and rinsed before refilling. Always inspect and test the unit if you have any doubt about its dependability. Identify problems or concerns and establish a regular maintenance program. Consult the manufacturer's operating manual and ANSI Z358.1 for assistance in performing test procedures, maintenance operations, and training. Personal eyewash bottles can provide immediate flushing when located in hazardous areas. However, personal eyewash equipment does not meet the requirements of plumbed or gravity-feed eyewash equipment. Personal eyewash units can support plumbed or gravity-feed eyewash units but cannot serve as a substitute (Table 6.2).

Chemical Storage Considerations

The OSHA Hazard Communication Standard (HCS) requires organizations to maintain a SDS for each hazardous substance used in the workplace. Refer to the SDS and in some situations the container label for information on special storage requirements. Typical storage considerations may include factors such as temperature, ignition control, ventilation, segregation, and identification. Properly segregate hazardous materials according to compatibility. For example, never store acids with bases or oxidizers with organic materials or reducing agents. Corrosives and acids will corrode most metal surfaces, including storage shelves or cabinets. Store flammable and combustible materials in appropriate rooms or approved cabinets.

Compressed Gas Safety

The Compressed Gas Association (CGA) promotes safe work practices for industrial gases and develops safe-handling guidelines. OSHA regulates the use and safety of compressed gases in the workplace. Refer to 29 CFR 1910.101 for complete information on inspecting gas cylinders. The DOT regulates the transportation of compressed gases by rail, highway, aircraft, and waterway. Store compressed gas cylinders in cool and dry areas that are well ventilated and meet fire-resistant standards. Never store compressed gas cylinders at temperatures higher than 125°F. Do not store cylinders near heat, open flames, or ignition sources. Properly label all cylinders, and never remove valve-protection caps until securing cylinder for use. Comply with OSHA 29 CFR 1910.101–105 and DOT 49 CFR 171–179 standards when handling compressed gases. Refer to ANSI-Z48.1 and CGA pamphlet C-7 for marking cylinders. When not in use, close valves and properly secure them. Use appropriate lifting devices to transport gas cylinders. Refer to the appropriate SDS for information about cylinder content. Inside of buildings, separate oxygen and flammable gas cylinders by a minimum of 20 ft. You can also store cylinders in areas with a fire-resistible partition between the oxygen and flammable materials.

OSHA Hazard Communication Standard (29 CFR 1910.1200)

OSHA requires employers to develop and implement a hazard communication plan. The plan must address container labeling, SDS availability, and training requirements. Employers must list the person responsible for each element of the plan. Organizations must make a copy of the written plan available upon request to employees on all shifts and to OSHA compliance officers. Communicate appropriate hazard information to all affected or exposed individuals. Chemical manufacturers and distributors must provide hazard information on their products through the means of container labels and SDSs.

Globally Harmonized System

GHS stands for globally harmonized system, the international approach to hazard communication. This global system provides criteria for classifying chemical hazards and standardizing labels and SDSs. Development of GHS required a multiyear endeavor by hazard-communication experts from different countries, international organizations, and stakeholder groups. OSHA recently modified the HCS to adopt the GHS approach. Since 1983, OSHA required employers to communicate hazardous materials information to employees. The original performance-oriented standard allowed chemical manufacturers and importers to convey information on labels and SDSs in a variety of formats. GHS requires a more standardized approach to classifying the hazards and conveying the information to those needing to know. The GHS requires providing detailed criteria for determining what hazardous effects a chemical poses. It also requires a standardized label assigned by hazard class and category. This will enhance both employer and worker comprehension of the hazards, resulting in safer use and handling. The harmonized format of the SDSs will enable employers, workers, health professionals, and emergency responders to access the information more efficiently and effectively. OSHA will require training on new label requirements and SDS format by December 2013. OSHA will require complete compliance in 2015.

Major Hazard Communication Standard Changes

The definitions of hazard will change to provide specific criteria for classification of health and physical hazards as well as classification of mixtures. These specific criteria will help to ensure that evaluations of hazardous effects remain consistent across manufacturers. This will result in more accurate labels and SDSs. Chemical manufacturers and importers must provide a label that includes a harmonized signal word, pictogram, and hazard statement for each hazard class and category. Precautionary statements must also be provided. Finally, the SDS will contain a specified 16-section format.

The GHS does not address harmonized training provisions. However, the revised Hazard Standard requires retraining of all workers within two years of the publication of the final rule. The parts of the OSHA standard not related to the new system—such as the basic framework, scope, and exemptions—remain unchanged. OSHA did modify some terms to align the revised standard with language used in the GHS. The term *hazard determination* has been changed to *hazard classification,* and *material SDS* was changed to *SDS.* Evaluation of chemical hazards must use available scientific evidence concerning such hazards. The revised standard contains specific criteria for each health and physical hazard, with instructions about hazard evaluations and determinations. The revised standard also establishes both hazard, classes and hazard categories. The standard divides the classes into categories that reflect the relative severity of the effect.

The original standard did not include categories for most of the health hazards covered. OSHA included general provisions for hazard classification and Appendixes A and B to address criteria for each health or physical effect.

Label Changes in the Revised Hazard Communication Standard

Under the original standard, the label preparer provided the identity of the chemical and the appropriate hazard warnings. The preparer determined the method to convey the information. The revised standard specifies what information to provide for each hazard class and category (Table 6.3).

The revised HCS requires printing of all red borders on the label with a symbol printed inside. Chemical manufacturers, importers, distributors, or employers who become aware of any significant information regarding the hazards of a chemical must revise labels on a chemical within 6 months of becoming aware of the new information. Employers may choose to label workplace containers with the same type of label affixed to the larger shipped containers. They can also use label alternatives that meet the requirements of the standard, including the National Fire Protection Association (NFPA) 704 Hazard Rating and the Hazardous Material Information System (HMIS). However, information supplied on alternative labels must meet requirements of the revised standard, with no conflicting hazard warnings or pictograms.

Safety Data Sheet Changes

The information required on the SDS remains essentially the same as the original standard. The original standard required specific information but did not specify a format for presentation or order of information. The revised standard requires presenting the information on the SDS using consistent headings in a specified sequence. The SDS format remains the same as the ANSI standard format (Table 6.4).

OSHA plans to retain the requirement to include the ACGIH TLVs on the SDS. OSHA finds that requiring TLVs on the SDS will provide employers and employees with useful information to help them assess the hazards presented by their workplaces. OSHA will also require the inclusion of PELs and any other exposure limits used or recommended by the chemical manufacturer,

Table 6.3 New Labeling Requirements

- Pictogram—This method uses a symbol plus other graphic elements, such as a border, background pattern, or color to convey specific information about the hazards of a chemical. Each pictogram consists of a different symbol on a white background within a red square frame set on a point (a red diamond). The system requires the use of nine pictograms. However, OSHA requires the use of only eight pictograms under the revised standard.

- Signal words—This requirement consists of using a single word to indicate the relative level of severity of hazard to alert the reader of a potential hazard. The signal words used include *danger* and *warning*. Use *danger* for severe hazard and *warning* for less-severe hazards.

- Hazard statement—This requirement consists of a statement assigned to a hazard class and category that describes the nature of the hazards of a chemical. It also includes, as appropriate, the degree of hazard.

- Precautionary statement—This phrase describes recommended measures to minimize or prevent adverse effects that could result from exposure to a hazardous chemical. It also applies to the improper storage or handling of a hazardous chemical.

Table 6.4 Required Safety Data Sheet Information

- Section 1. Identification
- Section 2. Hazard(s) identification
- Section 3. Composition/information on ingredients
- Section 4. First-aid measures
- Section 5. Fire-fighting measures
- Section 6. Accidental release measures
- Section 7. Handling and storage
- Section 8. Exposure controls/personal protection
- Section 9. Physical and chemical properties
- Section 10. Stability and reactivity
- Section 11. Toxicological information
- Section 12. Ecological information
- Section 13. Disposal considerations
- Section 14. Transport information
- Section 15. Regulatory information
- Section 16. Other information, including date of preparation or last revision

Note: OSHA does not mandate inclusion of Sections 12–15 in the SDS.

importer, or employer preparing the SDS. The revised standard provides classifiers with the option of relying on the classification listings of International Agency for Research on Cancer (IARC) and National Toxicology Program (NTP) to make classification decisions regarding carcinogenicity, rather than applying the criteria themselves.

OSHA also included a nonmandatory Appendix F in the revised standard to provide guidance on hazard classification for carcinogenicity. Part A of Appendix F includes background guidance provided by GHS based on the "Preamble" of the IARC "Monographs on the Evaluation of Carcinogenic Risks to Humans." Part B provides IARC classification information. Part C provides background guidance from the National NTP "Report on Carcinogens."

Addressing Pyrophoric Gases, Simple Asphyxiant, and Combustible Dust

The revised OSHA standard added pyrophoric gases, simple asphyxiants, and combustible dust to the definition of "hazardous chemical." OSHA also added definitions for pyrophoric gases and simple asphyxiants. It also provided guidance on how to define combustible dust for the purposes of complying with the standard. OSHA has retained the definition for pyrophoric gases from the original standard. OSHA requires addressing pyrophoric gases on container labels and SDSs. OSHA provided label elements for pyrophoric gases that include the signal word *danger* and the hazard statement "catches fire spontaneously if exposed to air." OSHA provided label elements for simple asphyxiants to include the signal word *warning* and the hazard statement "may displace oxygen and cause rapid suffocation." OSHA did not provide a definition for combustible dust but did require addressing dust on labels and SDSs. Label elements must include the signal word *warning* and the hazard statement "may form combustible dust concentrations in the air."

Managing and Communicating Changes to the Hazard Communication Standard

Consider the GHS as a living document with expectations of relevant updates on a 2-year cycle. OSHA anticipates future updates of the HCS to address minor terminology changes, final rule text clarification, and additional rulemaking efforts to address major changes.

Employee Training

The OSHA HCS (29 CFR 1910.1200) requires employers to provide employees information and training on hazardous chemicals used in their work areas. Employers must conduct training at the time of their initial assignment and upon the introduction of a new hazardous substance. Training must address the methods and observations used to detect the presence or release of the chemical. It must also address physical and health hazards, protective measures, labeling, and an explanation of the SDS. Employers must inform employees of the hazards of non-routine tasks and the hazards associated with chemicals in unlabeled pipes (Table 6.5).

Aspects of Pesticide Regulation That GHS Does Not Affect

Implementing GHS does not change most aspects of the pesticide program. It does not affect supplemental information required on labels, such as directions for use or additional hazard information that does not contradict or detract from GHS label requirements. It also does not impact testing methods for health and environmental hazards, data requirements, the scope of hazards covered, policies governing the protection of confidential business information, or risk-management measures.

Piping Systems Identification (ANSI Z13.1)

The standard addresses this concern by offering a common labeling method for use in all facilities. Ensure the labeling of pipes to communicate contents and provide information such as hazards, temperatures, and pressures. Use arrows to show the direction the material flows. High-hazard materials use black characters on a yellow background.

The low-hazard liquids or liquid mixtures use white characters on a green background. Low-hazard gases or gaseous mixtures use white characters on a blue background. Label fire-suppression pipes with white letters on a red background. Position labels for easy reading by placing them on the lower side of the pipe if the worker must look up or on the upper side of the pipe if workers must look down. Place labels near valves, branches, areas with direction changes, entry locations, and reentry points.

Table 6.5   HAZCOM-Mandated Training Topics

- Existence and requirements of the OSHA Hazard Communication Standard
- Components of the local hazard-communication program
- Work areas and operations using hazardous materials
- Location of the written hazard-evaluation procedures/communication program
- Location of the hazardous materials listing
- Location and accessibility of the SDS file

DOT Hazardous Material Regulations

The Hazardous Materials Table located in 49 CFR 172.101 provides the initial step toward understanding how to ship a product. This table provides the proper shipping name (PSN), hazard class, UN identification numbers, labels, and packaging types necessary. Locate the PSN from the alphabetically arranged Hazardous Materials Table (Table 6.6).

Identify the contents of a shipment using shipping papers, markings, labeling, and placard information. Refer to the Hazardous Materials Table (49 CFR 172.101). A marking can include handwriting or a preprinted, self-adhesive label containing required information such as PSN, the United Nations/North American (UN/NA) identification number, and the consignee's or consignor's name and address (49 CFR 172.300).

Labeling using a 4" × 4" square-on-point label helps visibly identify a hazardous materials package. Consider shipping labels as specific to the hazard classes of materials with strict specifications for setup, including color, size, and wording as well as placement on a package (49 CFR 172.400–172.450). The Hazardous Materials Table contains a label column referencing the label for the specific chemical by the hazard class. A label chart that shows hazard class or division and the associated label plus the section reference can be found in 49 CFR 172.400(b). When using two labels, the less hazardous of the two is a secondary hazard. This secondary hazard must contain labeling that meets requirements of 49 CFR 173.402. Use Special Precautions Labels such as "Empty" or "Cargo Aircraft Only" if required. OSHA Standard 29 CFR 1910.1201 requires original DOT labels to remain on vehicles, tanks, and containers until removal or transfer of labeled substances.

Placards

Depending on the nature and quantity of the shipment, placarding completes the shipment-identification process. Larger than labels, placards measure 10¾" × 10¾" but retain a similar square-on-point design. Placards deal with a specific hazard class of materials. DOT requires strict specifications for color, size, and wording and placement on a shipping vehicle (49 CFR 172.500.172.560). Two tables help determine the requirement for placards (49 CFR 172.504). DOT requires placards for secondary hazards. Secondary hazards must follow the requirements contained in 49 CFR 172.519.

Table 6.6 DOT Hazard Classes

- Class 1—Explosive (1.1–1.6)
- Class 2—Gases (2.1 flammable, 2.2 non-flammable, 2.3 poisons)
- Class 3—Flammable/combustible liquids
- Class 4—Solids (4.1 flammable, 4.2 spontaneously combustible, 4.3 dangerous when wet)
- Class 5—Oxidizing agents (5.1 oxidizers, 5.2 organic peroxide)
- Class 6—Poisons (6.1 poisons, 6.2 infectious material)
- Class 7—Radioactive (radioactive I, radioactive II, radioactive III)
- Class 8—Corrosive
- Class 9—Miscellaneous

Table 6.7 Hazardous Materials Training

- Specific requirement is located in 49 CFR172.704.
- Air 49 CFR 175.20, Vessel 49 CFR 176.13, and Highway 49 CFR 177.800–177.816.
- Use a systematic approach for training, testing, and documentation.
- Ensure employees understand the regulations and can perform assigned functions.
- 49 CFR 172.101, Hazardous Materials Table, designates what materials are hazardous.
- List encompasses a wide range of chemicals and combustible/flammable liquids.

Table 6.8 Hazardous Material Employees

- A person directly affecting hazardous materials transportation safety
- Owner or operator of a motor vehicle that transports hazardous materials
- A person loading, unloading, or handling hazardous materials
- A worker marking or taking other actions to qualify package transport
- A worker responsible for safety of transporting hazardous materials

Containers

Determining the applicable container for shipping a hazardous material depends on the UN identification code on the drum. For more information, refer to 49 CFR 178. To find information on a chemical, refer to the Hazardous Materials Table, SDS, Merck Index, CRC Handbook of Chemistry and Physics, or contact the manufacturer of the chemical (Tables 6.7 and 6.8).

Common Workplace Chemical Hazards

Workplaces can use a variety of solvents, such as methyl ethyl ketone, acetone, and Stoddard solvent. Workers who come into contact with solvents should wear recommended PPE. Most solvents can remove the natural fats and oils from the skin and some pose absorption risks. Organizations must store flammable solvents in approved containers. Provide local exhaust ventilation, and as needed use enclosures to control workplace exposures. When selecting appropriate engineering or other controls, safety personnel must consider the toxicity, flammability, and explosion potential of the material. Remove asbestos using only fully trained personnel adhering to the methods and protective equipment mandated by OSHA and EPA asbestos standards. OSHA 29 CFR 1910.1001 contains standards addressing working in or near in-place asbestos.

OSHA Construction Standard 29 CFR 1926.1101 covers activities such as demolition, removal, repair, or encapsulation of asbestos. Drain cleaning chemicals can burn the skin and damage the eyes. Workers should wear rubber gloves and goggles or face shields when they use drain cleaners if splashing is possible. Product information sheets or material SDSs contain additional information. Paints and adhesives contain a wide variety of solvents and should be used only in areas with adequate ventilation. When ventilation proves inadequate, workers should wear respirators approved for use with organic vapors. The EPA considers insecticides, herbicides, fungicides, disinfectants, rodenticides, and animal repellents as pesticides. They are considered to be hazardous substances under the OSHA HCS. EPA regulates pesticides under Federal Insecticide,

Fungicide, and Rodenticide Act regulations. Responsibility for the safe use of these toxic materials begins with purchase and continues until the empty container is properly discarded. All pesticides sold in the United States must carry an EPA registration number. The EPA Worker Protection Standard, 40 CFR 156 and 170, contains regulations that address the handling, loading, mixing, or applying of pesticides and the repairing of pesticide-applying equipment. Some pesticide products require verbal warnings and posted warning signs.

Managing Waste

Virtually all health-care or industrial facilities generate hazardous wastes as defined by the RCRA. Effectively managing inventory provides the next-best opportunity to reduce hazardous-waste generation. The program should be developed and updated as required. Never discard any hazardous chemical down the drain, in a toilet, or on the ground outside. Never attempt to burn chemical waste, under any circumstances. Never place hazardous chemicals in trash cans or garbage containers destined for landfills. Always read the label, check the SDS, and follow established facility procedures. Even a small amount of some chemicals, when left in a container, can pose a danger. Dispose of all waste containers according to required procedures. Wastes can react with one another and burn, release toxic vapors, or explode (Table 6.9).

The EPA requires generators to track all materials using the "cradle-to-grave" management approach. Develop policies and procedures for identifying, handling, storing, using, and disposing of hazardous wastes from generation to final disposal. Provide training for all exposed personnel. Monitor personnel who manage or regularly come into contact with hazardous materials and/or wastes. Evaluate the effectiveness of the program, and provide reports to senior leaders. RCRA Subtitle C regulations focus on the management of wastes with hazardous properties. The regulations help protect human health and the environment from mismanagement of hazardous wastes. Generators normally cannot store hazardous waste for more than 90 days, under EPA regulations. RCRA Subtitle C established four characteristics of hazardous waste. The four categories include corrosiveness, ignitability, reactivity, and toxicity. The EPA considers solid waste as hazardous waste if it meets the following criteria: (1) listed as a hazardous waste in the regulations, (2) substance contains a listed hazardous waste, and (3) waste was derived from the treatment, storage, or disposal of a listed hazardous waste.

EPA and state environmental agencies possess authority to inspect facilities and their records at any reasonable time. When an EPA inspector finds the facility in violation of RCRA or its permit, enforcement action in the form of compliance order, including administratively imposed injunctions or court actions, may follow.

Table 6.9 Waste Management Elements

- Inventory and categorization of wastes
- Identification hazards during daily activities
- Knowledge of storage requirements
- Distribution and special handling issues
- Special cleanup procedures
- Disposal procedures to include identification, transport, pickup, and end point

Other Types of Hazardous Waste

The EPA considers as hazardous any *ignitable solid waste* exhibiting any of the following properties: (1) any liquid, except aqueous solutions containing less that 24% alcohol, that has a flash point less than 140°F; (2) any nonliquid substance capable, under normal conditions, of spontaneous or sustained combustion; or (3) an ignitable compressed gas or oxidizing material as defined by DOT regulations. The EPA considers mixed waste as hazardous under RCRA and radioactive under the Atomic Energy Act. Both the NRC and EPA work together to address the management of these wastes. Generators of mixed waste must comply with both RCRA and NRC regulations.

Universal Waste

EPA's universal waste regulations streamline hazardous waste management standards for federally designated "universal wastes," which include batteries, pesticides, and mercury-containing equipment bulbs or lamps. Universal waste normally poses a relatively low risk during accumulation and transport. Recycling these wastes by adhering to the universal waste regulation facilitates environmentally sound management practices. The universal waste regulations ease the regulatory burden on retail stores and others wishing to collect these wastes. The regulations encourage the development of municipal and commercial programs to reduce the quantity of these wastes going to municipal solid waste landfills or combustors. In addition, the regulations also ensure that the wastes subject to this system will go to appropriate treatment or recycling facilities pursuant to the full hazardous waste regulatory controls. Refer to 40 CFR 273 for the federal universal waste regulations. States can modify the universal waste rule or add additional universal wastes in state regulations. The EPA recently proposed adding hazardous pharmaceutical waste to the universal waste rule.

Medical Waste

Health-care facilities, hospitals, clinics, physician offices, dental practices, blood banks, veterinary clinics, medical research facilities, and laboratories all generate medical waste. The Medical Waste Tracking Act of 1988 defines medical waste as "any solid waste that is generated in the diagnosis, treatment, or immunization of human beings or animals, in research pertaining thereto, or in the production or testing of biologicals." This definition includes, but is not limited to, blood-soaked bandages, culture dishes and other glassware, surgical gloves and instruments, sharps and needles, cultures, and pathological waste. The EPA also defines medical waste in 40 CFR 259.10 and 40 CFR 22 as any solid waste generated in the diagnosis, treatment, or immunization of human beings or animals. Most states regulate medical waste within their jurisdictions.

Electronic Waste (E-Waste)

Improperly disposing of electronic waste can result in risks to the environment and to human health. The use of electronic products continues to grow at a phenomenal pace. According to the Consumer Electronics Association, each American household contains about 24 electronic products. Donating used electronics for reuse extends the lives of valuable products. Recycling electronics prevents valuable materials from going into the waste stream. Many states currently have laws on the disposal and recycling of electronics. The EPA encourages all electronics recyclers to become certified and all customers to choose certified recyclers. Some electronic devices such as color cathode ray tubes (CRTs), computer monitors, color CRT TV tubes, cell phones, and other handheld devices could test as "hazardous" under federal laws. EPA encourages reuse and recycling of used electronics, including those that test as hazardous. The EPA does not consider computer monitors and televisions sent for continued use such as resale or donation as hazardous wastes. The EPA encourages recycling of CRTs, and those sent for recycling become subject to streamlined handling requirements. For more information on the CRT Rule, including export requirements, and frequent questions, please refer to the EPA Cathode Ray Tubes Final Rule.

Circuit boards receive a special exemption from federal hazardous waste rules. Whole used circuit boards meet the definition of spent materials but also meet the definition of scrap metal. EPA exempts recycled whole used circuit boards from hazardous waste regulation. EPA also excludes shredded circuit boards from the definition of solid waste if they are containerized before recovery. Shredded circuit boards cannot contain mercury switches, mercury relays, nickel cadmium batteries, or lithium batteries.

Refer to 40 CFR 260–262 for rules governing hazard waste generation. All facilities generating more than 100 kg or 220 pounds/month fall under federal hazardous waste regulations. EPA considers all CRTs coming from such facilities as permit-required hazardous waste. The EPA will exempt CRTs from businesses generating less than 100 kg or 220 pounds/month of hazardous waste. If a "small quantity generator" wishes to dispose of a small quantity of CRTs or other used electronics that does test hazardous under federal law, these materials can go to any disposal facility authorized to receive solid waste, unless state law requires more stringent management. State regulatory requirements for e-waste can be more stringent than the federal requirements and vary from state to state. Some states may develop universal waste exemptions for CRT. This action would streamline the management of CRTs bound for recycling. When planning to dispose of used CRTs or other electronics that test as hazardous under state or federal law, check state requirements to ensure compliance.

Hazardous Waste Operations and Emergency Response Standard

The Hazardous Waste Operations and Emergency Response standard contains requirements for cleanup operations and emergency-response operations for hazardous wastes. The standard (29 CFR 1910.120) requires that DOT-specified salvage drums or containers and suitable quantities of proper absorbent shall be kept available and used in areas where spills, leaks, or ruptures may occur. OSHA requires the development of a spill-containment program to contain and isolate the entire volume of the hazardous substance. Responders must meet the training requirements of the OSHA standard.

OSHA also mandates the use of appropriate PPE when responding to a spill or supporting decontamination activities. Responders must know the types of chemicals, level of exposure risk, and physical characteristics of the chemical hazard. Responders must know about potential hazards to the body. These hazards include toxicity, carcinogenic, asphyxiate, or corrosive nature of the chemicals. The EPA's Office of Emergency and Remedial Response defines four levels of protection for chemical-response operations. Level A is the highest level of skin and respiratory protection available. Protection must be appropriate for possible threats to life and health and during operations dealing with an unknown hazard. This level requires the highest level of respiratory protection with air-supplied respirators. Level B offers protection from chemical splash, but does not prevent exposure to gases or vapors. Level B requires the highest level of respiratory protection. Level C provides the same as level B but with an air-purifying respirator. Level D refers to the lowest level of protection and offers minimal protection for nuisance exposure. Refer to OSHA 29 CFR 1910.120 Appendix B for more specific information.

Chemical Protective Clothing

When selecting chemical-resistant clothing, consider materials measured by permeation testing methods of ASTM Standard F739. Breakthrough time refers to the time it takes the test chemical to pass from the outside surface of a clothing sample until detected on the inside surface of clothing. Permeation rate refers to the speed at which a chemical passes through the clothing once breakthrough has occurred. 29 CFR 1910, Subpart I, Appendix B recommends that for mixtures and formulated products (unless specific test data are available), select a glove with the shortest breakthrough time. Protective eyewear should match the work application. Goggles provide the most protection, since they form a seal around the eye area. Goggles come in vented and nonvented styles. Vented goggles offer protection from impact hazards only. Antifog lens can help increase vision. Face shields provide secondary protection against liquid splash, gases, vapors, or flying particles. Whenever a face shield is worn, primary protection such as goggles or safety glasses must also be worn.

Process Safety Management

OSHA published the Process Safety Management Standard (29 CFR 1910.119) in 1992. The standard applies to any process that contains a threshold quantity or greater amount of a toxic or reactive highly hazardous chemical (HHC) as specified in Appendix A. Also, it applies to 10,000 pounds or greater amounts of flammable liquids and gases and to the process activity of manufacturing explosives and pro-technics. The standard does not apply to retail facilities, normally unoccupied remote facilities, and oil or gas well drilling or servicing activities. The standard does not cover hydrocarbon fuels, used solely for workplace consumption, if not part of a process containing another HHC covered by the standard. Process safety management (PSM) does not apply to atmospheric tank storage and associated transfer of flammable liquids kept below their normal boiling point without benefit of chilling or refrigeration unless connected to a process or is sited in proximity to a covered process. The OSHA standard requires compilation of written process safety-related policies that must include documentation information on HHCs. OSHA requires development of a written plan of action that addresses employee participation. The plan must include consulting with employees on conducting and developing process hazard analyses and other elements of PSM. Employees include both work site and contractor employees. The standard requires the completion of the hazard analyses as soon as possible for each covered process. Update the process and revalidate every 5 years during the life of the process.

Process safety consists of blending engineering and management skills focused on preventing catastrophic accidents such as explosions, fires, and toxic releases associated with the use of chemicals and petroleum products. Many facilities must comply with the OSHA standard and the EPA Risk Management Program (RMP) standards published in 40 CFR Part 68. The EPA model RMP plan for an ammonia refrigeration facility provides excellent guidance on how to also comply with both standards. The Center for Chemical Process Safety of the American Institute of Chemical Engineers published a widely used book that explains various methods for identifying hazards in industrial facilities and quantifying their potential severity.

Table 6.10   Safety from External Exposure Sources

> • Time—decrease the amount of time you spend near the source of radiation.
>
> • Distance—increase the distance from a radiation source.
>
> • Shielding—increase the shielding between you and the radiation source.

Ionizing Radiation

Ionizing radiation hazards exist in many organizations, including health-care facilities, research institutions, nuclear reactors and their support facilities, nuclear weapon production facilities, and other various manufacturing settings. OSHA addresses ionizing radiation in standards for the general industry, shipyard employment, and the construction industry. Radiation includes alpha rays, beta rays, gamma rays, X-rays, neutrons, high-speed electrons, high-speed protons, and other atomic particles. The term does not include sound or radio waves, visible light, or infrared (IR) or ultraviolet (UV) light. Radioactive material means any material that emits, by spontaneous nuclear disintegration, corpuscular or electromagnetic emanations (Table 6.10).

Nuclear Regulatory Commission

The NRC ensures civilian uses of nuclear substances meet the requirements of safety, health, environmental, and national security laws. The NRC issues a 5-year license to qualified organizations that follow prescribed safety precautions and standards. Each NRC licensee must develop a written management plan to direct the active participation of users, administrators, and radiation safety officers. Organizations must also establish procedures to address the types and amounts of materials used, dosing information, safety precautions, recurring training, and continuing education. Organizations must develop procedures to investigate all incidents, accidents, or other deviations from prescribed procedures. Work with the radiation safety committee in overseeing the facility radiation-safety program.

Ionizing Exposures

Ionizing radiation comes from the natural decay of radioactive materials or through X-ray-producing devices. A radioactive element can spontaneously change to a lower energy state and emit alpha particles, beta particles, or gamma rays. X-rays come from highly energized electrons striking the nuclei of the target material. The electrons deflect from their path and then release energy in the form of electromagnetic radiation, or X-rays. Ionizing radiation hazards vary in the ability to penetrate the body and produce harmful effects. The ability to penetrate the body depends on the wavelength, frequency, and energy of the material. Alpha particles consist of two neutrons and two protons. These particles do not penetrate the skin. A thin layer or paper can shield the skin. Beta particles can travel a few centimeters into tissue. These particles can travel short distances in the air and can moderately penetrate the body. If beta-emitting contaminants remain on the skin for a prolonged period of time, they may cause skin injury. Beta-emitting contaminants *may cause harm* if deposited internally. Gamma rays and X-rays penetrate human tissue and can cause damage. Radioactive materials that emit gamma radiation can become an external and internal hazard to humans. Use dense materials to shield against gamma radiation. PPE provides little shielding from gamma radiation but will prevent contamination of the skin. Although similar to gamma rays, X-rays possess longer wavelengths, lower frequencies, and lower energies.

Agreement State Program

NRC can enter into agreements with states desiring to assume regulatory responsibility under Section 274 of the Atomic Energy Act of 1954, as amended. The NRC can relinquish to states portions of its regulatory authority to license and regulate by materials or radioisotopes, source materials such as uranium and thorium, and certain quantities of special nuclear materials. The mechanism for transferring NRC authority to a state occurs after the governor of the state and the chairman of the commission sign an agreement. NRC conducts training courses and workshops, evaluates technical licensing, and assesses inspection issues from agreement states.

Units of Measure

Most scientists in the international community measure radiation by using System International (SI) based on a metric system. The United States uses a conventional system of measurement. Different units of measure vary, depending on what aspect of radiation is being measured. The measurement of the amount of radiation given off or emitted uses the conventional unit curie (Ci). We can measure radiation dose absorbed by using the conventional unit rad or the SI unit referred to as gray (Gy). The biological risk of exposure to radiation uses the conventional unit called rem or the sievert (Sv). One sievert equals 100 rem. About 80% of typical human exposure comes from naturally occurring sources, and the remaining 20% comes from artificial radiation sources such as medical X-rays.

National Council on Radiation Protection

Congress created National Council on Radiation Protection (NCRP) to collect, analyze, develop, and disseminate information and recommendations on radiation quantities, measurements, and units. NCRP publishes maximum exposure permissible levels of external and internal radiation. The major handbook sources include *Maximum Permissible Body Burdens and Maximum Permissible Concentrations of Radionuclides in Air and in Water for Occupational Exposure* and *Review of the Current State of Radiation Protection Philosophy*. The NCRP suggests an annual permissible whole body dose of 5 rem/year, with 3 rem permitted within a 13-week period. NCRP goals are primarily to prevent and reduce cataracts, erythematic conditions, and probability of cancer.

Food and Drug Administration

The Food and Drug Administration (FDA), under the Federal Food, Drug, and Cosmetic Act, has the authority to regulate the manufacture and distribution of radiopharmaceuticals and medical devices that contain radioactive materials. The FDA also sets performance standards for X-ray and other radiation-emitting equipment that were manufactured after 1974. The FDA also issues recommendations for the use of X-ray machines and other radiation-emitting devices. Title 10 CFR

Parts 20 and 34 contain NRC rules on isotope sources. FDA X-ray regulations are found in Title 21 CFR Parts 1000 and 1050.

OSHA Ionizing Radiation Standard (29 CFR 1910.1096)

OSHA regulates exposure to all ionizing radiation from sources not under NRC jurisdiction. OSHA standards cover X-ray equipment, accelerators, accelerator-produced materials, electron microscopes, and naturally occurring radioactive materials such as radium. An OSHA and NRC Memorandum Agreement of 1989 outlined compliance authority of both agencies. A coordinated interagency effort can help prevent gaps in the protection of workers and avoid duplication of effort. OSHA conducts inspections at NRC-licensed nuclear power plants when accidents, fatalities, referrals, or worker complaints occur. OSHA can report the following to the NRC: (1) poor security control or work practices that would affect radiological safety, (2) improper posting of radiation areas, and (3) licensee employee allegations of NRC license or regulation violations. The OSHA exposure standards for whole-body radiation must never exceed 3 rem/quarter (year). Lifetime or cumulative exposure must never exceed $5(N - 18)$ rem. OSHA defines a "radiation area" as location accessible to personnel, in which there exists radiation at such levels that a major portion of the body could receive in any single hour a dose in excess of 5 mrem or in any 5 consecutive days a dose exceeding 100 mrem. Caution signs, labels, and signals prescribed by OSHA must use the conventional radiation caution colors of magenta or purple on yellow background. The symbol prescribed consists of a conventional three-bladed design with the words *radiation area*. Each radiation area must contain conspicuously posted signs bearing the radiation caution symbol. Each area or room containing radioactive material other than natural uranium or thorium in any amount exceeding 10 times the quantity of such material specified in Appendix C to 10 CFR 20 must contain conspicuously posted signs that bear the radiation caution symbol and the word "Caution". Properly secure all radioactive materials stored in non-radiation areas to prevent unauthorized removal. No employer shall dispose of radioactive material except by transfer to an authorized recipient or in a manner approved by the NRC or similar state-level agency.

Radioactive Waste Management

The NCRP issues recommendations for dealing with radioactive wastes. Wastes can exist in the form of solids, liquids, or gases. Solid wastes can include rags and papers from cleanup operations, solid chemicals, contaminated equipment, experimental animal carcasses, and human or experimental animal fecal matter. The major ways to manage waste products include dilution, containment, incineration, and return to supplier. The properties of the material must be considered in the method of disposal. Specific disposal methods vary according to the material involved and the licensing authority of the user. Consider half-life and relative biological hazards when dealing with radioactive waste. Disposal can depend on the half-life of the radionuclide. Dispose of material with a short half-life by contracting with a commercial contractor. Retain some waste, such as that generated by departments of nuclear medicine and clinical testing laboratories, on-site, until their half-life expires. Segregate these low-level wastes by considering the isotope, form, volume, laboratory origin, activity, and chemical composition. Ensure the proper labeling of all wastes. Never mix radioactive with other types of hazardous wastes.

OSHA Nonionizing Radiation Standard (29 CFR 1910.97)

Electromagnetic radiation can produce varying effects on humans, depending on the wavelength and type of radiation involved. Consider low-frequency radiation, such as that generated by broadcast radio, as not dangerous. Thermal heating remains the greatest hazard associated with exposure to microwave (MW) radiation. The OSHA voluntary exposure limit for MW (10 mW/cm²) remains unenforceable. Some states that operate their own OSHA programs may enforce the limit or other established radio frequency (RF) exposure limits. The OSHA Construction Standard does specify the design of an RF warning sign. The OSHA Construction Standard 29 CFR 1926.54 does limit worker exposure to 10 mW/cm², which includes the painting of towers. The American Industrial Hygiene Association provides detailed information on the physical characteristics of RF and MW radiation, including generation and sources, how it interacts with matter, and its biological effects. Nonionizing radiation is described as a series of energy waves composed of oscillating electric and magnetic fields traveling at the speed of light. Nonionizing radiation includes the spectrum of UV, visible light, IR, MW, RF, and extremely low frequency. Lasers commonly operate in the UV, visible, and IR frequencies. Nonionizing radiation occurs in a wide range of occupational settings and can pose health risks to potentially exposed workers if not properly controlled. *Nonionizing radiation* refers to any type of electromagnetic radiation that does not carry enough energy per quantum to ionize atoms or molecules and removes electrons from an atom or molecule. No specific OSHA standards exist for RF and MW radiation issues. Other standards related to nonionizing risks include 29 CFR 1910.147, Control of Hazardous Energy and 29 CFR 1910.268, Telecommunications.

Standards for RF Exposures and Measurements

The Institute of Electrical and Electronics Engineers (IEEE) International Committee on Electromagnetic Safety sets safety standards for frequencies from 0 to 300 GHz. Also, the IEEE Committee on Man and Radiation (COMAR) publishes position papers on human exposure to electromagnetic fields. The ACGIH's *Documentation of the Threshold Limit Values for Physical Agents*, 7th edition, provides consensus exposure limits from organization of governmental industrial hygienists for RF and MW radiations. The FCC updated its RF safety regulations in 1997, which apply to transmitting sites in the United States. Individual states must meet all aspects of these regulations. Regulations set limits for human exposure. The FCC publishes limits that address spatial exposures averaged over the whole body. Occupational/controlled limits use a time average of over 6 minutes. The general population/uncontrolled exposure maximum permissible exposure (MPE) limits do not use time averaging. Although the FCC announced plans to issue exposure limits for induced and contact currents, the agency did not take any further regulatory actions. The FCC faces difficulty in making the measurements, due to lack of suitable equipment. Seek to reduce RF exposures through implementation of appropriate, administrative, work practice, and engineering controls. Specific absorption rate provides a common measure for RF exposure or the rate of energy absorption in tissue. It is measured in W/kg of tissue.

Ultraviolet Radiation

UV light can cause burns to skin and cataracts to the eyes. UV classifications can include near, medium, and far UV energy. Consider near UV radiation as nonionizing. UV light produces free radicals that induce cellular damage, which can be carcinogenic. UV light also induces melanin production from cells to cause sun-tanning of skin. Plastic sunglasses using polycarbonate generally absorb UV radiation. UV overexposure to the eyes causes snow blindness when at sea or in areas with snow on the ground.

Visible and Infrared

Visible light causes few effects to the human body. Bright visible light irritates the eyes. Visible-light lasers have much more powerful effects and may damage the eyes even at small powers. Very strong visible light can cauterize hair follicles. Electric and magnetic physical agents contain the potential to cause adverse *human effects*. General health effects reviews explore possible carcinogenic, reproductive, and neurological effects. Research continues on health effects of exposures to devices such as traffic radar, wireless communication devices, and magnetic resonance imaging (MRI). *Laser* is an acronym for to Light Amplification by Stimulated Emission of Radiation.

The energy generated by the laser occurs in or near the optical portion of the electromagnetic spectrum. Amplified energy of atomic processes results in an event called stimulated emission. Some people often misinterpret the term *radiation* when used with lasers. *Radiation* simply refers to an energy transfer. Energy moves from one location to another by conduction, convection, or radiation. We can express color of laser light in terms called wavelength. A nanometer expresses a laser's wavelength with billion nanometers in 1 m. Nonionizing laser light ranges from the UV (100–400 nm) to, visible (400–700 nm) to IR (700 nm–1 mm). A laser produces an intense and highly directional beam of light. Laser light, when directed, reflected, or focused, becomes partially absorbed by the object in its path. This reaction raises the temperature of the surface and/or the interior of the object. This can result in alteration or deformation of material, such as eye damage.

Laser Standards

The ANSI oversees the promulgation of the ANSI-Z136 series of laser safety standards. These standards provide the foundation of laser safety in industry, medicine, research, and government. OSHA evaluates laser-related occupational safety issues by referencing ANSI Z-136 laser safety standards. The Laser Institute of America (LIA) serves as the international society for laser applications and safety. LIA remains committed to the mission of fostering lasers, laser applications, and laser safety worldwide. The FDA regulates product performance for all laser products sold in the United States. The manufacturer must meet FDA performance and safety standards to certify products. Each laser must bear a label indicating compliance with the FDA standards and include the laser hazard classification.

ANSI Z-136.1, Safe Use of Lasers

This standard provides information on how to classify lasers for safety, laser safety calculations and measurements, and laser hazard control measures. The standard also addresses requirements for laser safety officers (LSOs) and laser safety committees. The standard specifies that design of signs and labels be in accordance with ANSI Z535 series of standards for accident-preventing signs.

The ANSI Z535 standard specifies sign dimension, lettering size, color, and other important sign-design elements. The new hazard signs contain an equilateral triangle attention symbol in addition to the familiar sunburst pattern. The new triangular symbol is introduced into both the caution and danger signs. As before, the International Electrotechnical Commission (IEC) signs and labels are specified as an acceptable alternative to the ANSI signs. The standard possesses classification guidelines and requirements closely harmonized with the corresponding international laser safety standard issued by the IEC.

ANSI Z-136.3, Safe Use of Lasers in Health Care

The standard serves as the definitive document on laser safety for all health-care environments. It provides guidance for the safe use of lasers for medicine, diagnostic, cosmetic, preventative, and therapeutic applications in any location where bodily structure or function is altered or symptoms are relieved. Lasers used in these applications are incorporated into an apparatus, which includes a delivery system, a power supply, mechanical housing, and associated liquids and gases as required for operation of the laser. The widespread use of medical lasers such as those used in ophthalmic refractive surgery and for various dermatological procedures moved lasers into conventional medical, surgical, and allied professions. The majority of medical laser systems now operate in private medical offices. The rapid change in usage patterns made revision of the standard essential.

ANSI Z136.4, Recommended Practice for Laser Safety Measurements

This standard provides guidance for optical measurements associated with laser safety requirements. The information contained in this document will help users who conduct hazard evaluations and ensure the use of appropriate control measures. It contains clearly written definitions, examples, and other practical information for manufacturers, LSOs, technicians, and other trained laser users.

ANSI Z136.5, Safe Use of Lasers in Educational Institutions

The standard applies the requirements of the ANSI Z136.1 to the unique environments associated with educational institutions. Such settings include teaching laboratories, classrooms, lecture halls, science fairs, and science museums that use lasers in their educational process. The standard applies to staff and students using lasers for academic instruction in university, college, secondary, or primary educational facilities.

ANSI Z136.6, Safe Use of Lasers Outdoors

This standard provides guidance for the safe use of lasers in an outdoor environment. It covers product performance of lasers used outdoors, including those with variances or exemptions from the provisions of the federal product performance standard (21 CFR 1040).

ANSI Z136.7, Testing and Labeling of Laser Protective Equipment

The standard provides reasonable and adequate guidance on the test methods, protocols, and specifications for devices used to provide eye protection from lasers and laser systems.

ANSI Z136.8, Safe Use of Lasers in Research, Development, or Testing

This standard addresses laser usage in labs and other research-designated areas. The standard details different laser-use locations as well as noting two additional hazard analysis areas—beam path and beam interaction. Highlights of this standard include the use of alignment eyewear, use of noncertified lasers, export controls, and use of warning signs. The standard contains sample audit forms for lab and program reviews.

Laser Hazard Classification

Research studies, along with knowledge of hazards related to sunlight and conventional man-made light sources, permitted development of safe exposure limits for nearly all types of laser radiation. Establishing limits referred to as MPE led to development of a system of laser hazard categories or classifications.

The FDA classifies lasers on potential to cause injury. The most relevant parameters used to classify a laser are the laser output energy or power and wavelengths. Refer to ANSI Z136.1, Section 3, for information on laser classifications.

Class 1 Lasers

Consider all Class 1 lasers as incapable of producing damaging radiation levels. Therefore, consider the laser safe under normal working conditions. Most control measures do not apply to these lasers. Many lasers in this class operate from an imbedded enclosure that prohibits or limits access to the laser radiation.

Class 2 Lasers

Class 2 lasers operate at low power, with output not exceeding 1 mW. The normal human aversion response of 0.25 seconds to bright radiant sources affords eye protection if the beam is viewed directly. A potential for eye hazard does exist for slower reflex time or exposure time greater than 0.25 seconds.

Class 1M Lasers

Class 1M lasers do produce hazardous exposure conditions during normal operation, unless the beam view occurs using an optical instrument such as an eye-loupe or a telescope.

Class 2M Lasers

Class 2M lasers emit visible radiation in the 400–700 nm range but with power output below 1 mW. These lasers pose little ocular hazard to the unaided eye but become potentially hazardous when viewed with optical aids.

Class 3R Lasers

Class 3R lasers pose some hazards under some direct and specular reflection viewing conditions with a low probability of an injury. Class 3R lasers do not pose either a fire or diffuse-reflection hazard. Class 3R lasers produce an output between one and five times the Class 1 power limit for wavelengths shorter than 400 nm (UV lasers), longer than 700 nm (IR lasers), or a output power of 5 mW for 400–700 nm wavelengths (visible lasers).

Class 3B Lasers

Class 3B lasers produce medium power, with output power of 5–500 mW. Never permit the viewing of these hazardous lasers under direct-beam or specular-reflection conditions. The diffuse reflection usually does not pose a threat except for higher-power Class 3B lasers. A Class 3B laser does not pose a fire hazard.

Class 4 Lasers

Class 4 lasers produce high-power output, above 500 mW. Exposure to the direct beam, specular reflections, or diffuse reflections presents a hazard to both the eyes and skin. A Class 4 laser poses a fire hazard risk (radiant power >2 W/cm^2 is an ignition hazard). These lasers can create hazardous airborne contaminants. They also pose a high-voltage electrical risk. Always enclose the entire laser beam path, if possible, or enclose most of the beam path to reduce the potential hazards.

Laser Safety Officers

ANSI Z136.1 specifies that any facility using Class 3b or Class 4 lasers or laser systems should designate a LSO to oversee safety for all operation, maintenance, and servicing situations. This person should have the authority and responsibility to monitor and enforce the control of laser hazards. This person must evaluate laser hazards and establish appropriate control measures. Some of the other duties of the LSO can include dealing with compliance issues, approving standard operating procedures, overseeing maintenance/service procedures, and supervising safety training.

Nanotechnology

The National Nanotechnology Initiative (NNI) defines nano technology as understanding and controlling of matter at dimensions of roughly 1–100 nm. In theory, these materials can be engineered from almost any material. Nanoscale materials appear in industrial and consumer products, including new drug-delivery formulations. Very little research exists on the potential toxicity of manufactured nanoscale materials.

The unique and diverse physicochemical properties of nanoscale substances suggest that toxicological properties may differ from materials of similar composition but in larger sizes. Nanotechnology can impact medical, ethical, mental, legal, and environmental fields. They also impact areas such as engineering, biology, chemistry, computing, materials science, military applications, and communications.

Benefits of nanotechnology include improved manufacturing methods, water purification systems, energy systems, physical enhancement, nanomedicine, and better food-production methods. Risks include environmental, health, and safety issues. When materials are engineered at the scale of atoms and molecules, they can behave in unconventional ways. An increasing body of research indicates that some of these materials may pose harm, but information remains scarce on what constitutes "due care" in many situations. The Academies of Science called for a national research strategy for nanotechnology risk research. The debate continues about the role that special government regulation and compliance agencies such as OSHA, EPA, and Consumer Products Safety Commission play in the safety of nanotechnologies. Group potential risks fall into one of four categories: health issues; environmental issues; societal concerns; and effects on politics, including human interactions and risks associated with the speculative vision of molecular nanotechnology. Two major safety concerns exist. The first deals with nanocomposites with nanostructured surfaces. The second concerns nanocomponents such as electronics, optical devices, and sensors. Nanoscale particles can incorporate into a substance, material, or device and become fixed particles. Free nanoparticles used in production of a substance could become a nanoscale species of elements or simple compounds. Many experts express a concern about the safety of free nanoparticles, because potential adverse effects cannot be derived from the known toxicity of the macro-sized material. Nanotechnology's health implications can be split into two aspects. The first concern addresses the potential for nanotechnological innovations to cure disease during medical applications. The second concern deals with potential health hazards posed by exposure to nanomaterials.

NIOSH Research

NIOSH continues researching technologies to evaluate the toxicological properties of major nanoscale material classes. The materials represent a cross section of composition, size, surface coatings, and physicochemical properties. We do not know fully how nanoscale materials can interact with biological systems. NIOSH recently published *Approaches to Safe Nanotechnology: Managing the Health and Safety Concerns Associated with Engineered Nanomaterials*, DHHS (NIOSH) Publication 2009-125. This document reviews current knowledge about nanoparticle toxicity, process emissions, exposure assessment, engineering controls, and PPE. This update incorporates the latest results of NIOSH research. NIOSH recommendations in *Current Intelligence Bulletin 60: Interim Guidance for the Medical Screening and Hazard Surveillance for Workers Potentially Exposed to Engineered Nanoparticles* responds to employers' ongoing interest in having authoritative occupational safety guidance for the manufacturing and industrial use of engineered nanomaterials.

Review Exercises

1. List at least five issues to consider when evaluating hazardous material risks.
2. How does OSHA define a hazardous substance?
3. What does the EPA add to the OSHA definition?
4. How does the DOT define a hazardous substance?
5. Define the following acronyms:
   PEL
   STEL
   REL
   TLV
6. Explain in your own words the purpose of the OSHA additive formula found in 29 CFR 1910.1000.
7. List the seven elements required by OSHA in an organization's respirator plan.
8. Describe the following respirators:
   Air-purifying respirator
   Supplied air respirator
   Self-contained breathing
9. Explain the difference between QNFT and a QLFT of a respirator.
10. List at least five physical properties of a chemical substance.
11. What two organizations work to publish consensus eyewash and shower standards?

12. List at least four of the basic requirements for eyewash and shower facilities.
13. What is the foundational consideration when storing hazardous substances?
14. What two regulatory agencies have enforcement authority related to storing compressed-gas cylinders? Describe the authority of each agency.
15. List and describe the elements in the new GHS labeling system.
16. List at least five HAZCOM training topics mandated by OSHA.
17. List the nine primary DOT hazard classes.
18. Which federal agency regulates pesticides, including registered disinfectants?
19. Describe the best three protections against exposure to ionizing radiation.
20. How does the NRC define an agreement state?
21. What role does the FDA play in radiation safety?
22. List five types of nonionizing radiation.
23. List and describe the ANSI-defined laser classes.

Chapter 7

Fire Safety and Emergency Management

Introduction

Hazard control managers must possess an understanding of basic fire safety principles and emergency management concepts. This chapter provides an overview of helpful information related to fire prevention safety and emergency management. The lack of effective emergency action plans or fire prevention plans could lead to severe losses, including casualties and even financial collapse of an organization. Emergency planning and fire safety remains an important aspect of organizational operation. Planning and then executing the fire plan or emergency operations plan (EOP) can save time, resources, and lives. Effective emergency and fire safety planning provide guidance for handling unexpected situations. Comprehensive emergency planning must consider all hazards, all phases, all stakeholders, and all impacts relevant to any emergency situation or disaster. This all hazards approach requires operational managers to anticipate potential events and take necessary preparatory measures.

Occupational Safety and Health Administration (OSHA) requires covered entities to develop a written fire prevention plan. Employers must maintain the plan in the workplace and make it available for employee review. However, an employer with 10 or fewer employees may communicate the plan requirements orally to the employees. A fire prevention plan must include a list of all major fire hazards, proper handling and storage procedures for hazardous materials, potential ignition sources and their control, and the type of fire protection equipment necessary to control each major hazard. The plan must also detail procedures for controlling accumulations of flammable and combustible waste materials. The plan must outline procedures used to ensure the regular maintenance of safeguards installed on heat-producing equipment to prevent the accidental ignition of combustible materials. The plan must contain the name or job title of employees responsible for maintaining equipment to prevent or control sources of ignition or fires. It also must contain the name or job title of employees responsible for the control of fuel source hazards. Employers must inform employees on initial job assignment information about fire hazard exposures in the workplace. The employer must also review with each employee all information in the fire prevention plan that addresses self-protection.

Life Safety Code, National Fire Protection Association 101

The life safety concept began in 1963 with the publication of the Building Exits Code. National Fire Protection Association (NFPA) published the first edition of the Life Safety Code® in 1966. Building codes provide guidance for overall design criteria. The Life Safety Code, NFPA 101®, addresses the general requirements for fire protection and systems safety necessary to assure the safety of building occupants during a fire. The code specifies minimum hourly fire-resistance ratings and does not specify how such ratings are to be achieved. Organizations should follow the appropriate edition of NFPA 101. Managers should determine occupancy type and apply the standard appropriately. When planning egress, develop procedures to ensure safe access and evacuation for all disabled individuals. Develop procedures to ensure fire safety during all phases of a renovation or construction project. Maintain and install fire protection equipment that meets NFPA codes and standard requirements. Provide necessary education and training to reduce fire-related risks and hazards. Evaluate safety training annually to ensure everyone understands expectations. Remember, NFPA 101 provides minimum requirements for the design, operation, and maintenance of buildings and structures. The code helps organizations ensure safety to life from fire and similar emergencies. NFPA 101 requires that new and existing buildings allow for prompt escape or provide people with a reasonable degree of safety through other means. It defines hazards and addresses general requirements for egress and fire protection features such as "fire" doors. The code also addresses building service and fire equipment such as heating components, ventilation, air-conditioning systems, sprinkler systems, fire detection systems, and localized extinguishers. Provisions can vary, depending on the type of occupancy and building type (new or an existing construction). The code can stand alone or be used in concert with other building codes, depending on the jurisdiction. New editions of the code build on the prior editions.

Understanding Fire

Fire creates a very rapid chemical reaction between oxygen and a combustible material. This reaction results in the release of heat, light, flame, and smoke. Fire needs a sufficient amount of oxygen to sustain combustion and a sufficient amount of heat to raise the material to its ignition temperature. Fire also needs some type of fuel or combustible material to burn (Table 7.1).

Table 7.1 Fire Development Stages

1. Incipient stage—during this stage, no smoke, flame, or significant heat buildup is visible, but a large quantity of combustion particles created by chemical decomposition can rise to the ceiling.
2. Smoldering stage—as the incipient stage continues, the combustion particles increase until they become as "smoke" with no flame or significant heat source.
3. Flame stage —as the fire condition develops further, ignition occurs and flames appear, with the level of visible smoke decreasing.
4. Heat stage—at this point, large amounts of heat, flame, smoke, and even toxic gases become apparent. This stage develops very quickly, usually in seconds.

Deficiencies

Fire safety deficiencies are on building-by-building basis and rated according to how deficiency affects overall fire safety. Deficiencies are classified as Level I, II, and III deficiencies.

■ Level I: a deficiency or series of deficiencies that indicate a lack of proper maintenance of building components that play a role in the unit concept.
■ Level II: a deficiency or series of deficiencies involving one or more unit-concept units that poses a threat to life. The scope of the deficiencies is significant in a limited area.
■ Level III: a deficiency or deficiencies indicating pervasive violation of one or more of the unit-concept units. The scope of the deficiencies is such that correction in less than three years is not possible.

Fire Safety Evaluation System

The National Bureau of Standards' Center for Fire Research promotes the use of a system for determining how combinations of fire safety elements can meet the intent of NFPA 101®. For example, the quantitative evaluation system grades fire safety using a "zone-by-zone" approach in health-care facilities. The system requires scoring the following areas: (1) containment actions taken to control the fire and smoke, (2) extinguishing systems and procedures to effectively extinguish a fire, (3) the process of moving occupants to safety from a fire or smoke zone, and (4) general safety procedures and policies that affect the overall safety of the fire and/or smoke zone.

Design Considerations

Facilities must be designed, constructed, maintained, and operated to minimize the possibility of a fire requiring the evacuation of occupants. Organizations must develop operating and maintenance procedures to address design, construction, and compartmentalization. Install appropriate detection, alarm, and extinguishing systems. Provide fire prevention training to include procedures for isolating fires. Make plans for total building evacuation or the use of refuge areas.

Fire Prevention

Publish specific fire prevention and protection measures. Buildings must operate a fire alarm or fire detection system that activates an alarm automatically in the event of a fire. Organizations must install air conditioning and ducts to meet the requirements of NFPA 90A, *Standard for Installation of Air Conditioning and Ventilating Systems*. Ensure all alarms generate a distinct sound loud enough for occupants to distinguish over normal operational noise levels. Locate manual fire alarm stations near each exit. Connect all electrical monitoring devices of automatic sprinklers to an alarm. Ensure fire extinguisher placement considers the type of fire probable for an area. Inspect fire extinguishers at least monthly, and ensure regular maintenance. Test fire alarms and detection systems once each quarter. Publish a smoking policy and implement appropriate electrical safety policies. Educate all personnel about fire safety and response plans.

Conduct quarterly fire inspections with an emphasis on the following:

■ Evaluation of all equipment
■ Testing of alarms, detectors, and pull stations
■ Evaluation of housekeeping practices and security alarms
■ Inspection of sprinkler system pressure
■ Ensuring of water availability and hydrant operation
■ Checking of suppression, detection, and activation systems annually

General Alarm Requirements

Fire alarms must signal a central location within the facility, and facilities must ensure the staffing of all supervised locations continuously. Designate and protect any supervised location as a hazardous area. Signals received must transmit immediately to the local fire department. Position manual alarm stations to ensure travel of no more than 200 ft. to reach an alarm station on the same floor. Audible alarm sounds must exceed the level of any operational noise. All visible alarms should also generate an audible alarm sound. Monitor electronic components to ensure awareness of any systems needing repair. The system should signal trouble when a break or ground fault interrupter prohibits normal system operation because of main power source failure or a break in the circuit wiring.

Fire System Inspections

All components of a fire alarm system must receive regular inspection and testing. All systems should receive a visual inspection each quarter. Each automatic system should be tested and/or inspected on an annual basis. Ensure inclusion of fire components and systems in an organizational preventive maintenance plan. Ensure the testing of all supervisory signal devices on a quarterly basis. Test valve tamper switches, water flow devices, duct detectors, electromechanical releasing devices, heat detectors, manual fire alarm boxes, and smoke detectors on a semiannual basis. Test occupant alarm notification devices to include audible and visible devices at least annually. Maintain appropriate testing documentation on all fire-related systems.

Fire Confinement

Confinement measures include dividing a building in a way to break up the total volume of a building into small cells. To assure proper protection of openings, install fire doors in accordance with NFPA 80, Standard for Fire Doors and Fire Windows. Many factors can determine the movement of smoke within a structure, including building height, ceiling heights, suspended ceilings, venting, external wind force, and the direction of the wind. One method of controlling smoke uses a physical barrier such as a door or damper to block the smoke's movement within the compartment. Regardless of the type of building construction, stair enclosures remain vital to safe exit for occupants. They also retard the upward spread of fire.

An alternative method uses a pressure differential between the smoke-filled area and the protected area. Venting provides another way to remove smoke, heat, and gases from a building. Vents can prove useful in windowless and underground buildings.

Emergency Exits

Designing exits involves more than a study of numbers, flow rate, and population densities. Safe exits also require a safe path of escape from the fire. Exit doors must withstand fire/smoke during the length of time for which it is designed. Provide alternative exits and pathways, in case one exit is blocked by fire. Provide exits with adequate lighting, and mark exits with readily visible signs. Protect exiting personnel and hazard areas that could contribute to the spread of fire and smoke. Develop plans to evacuate disabled or wheelchair employees to meet OSHA requirements. The American Disabilities Act (ADA) Title III requires most organizations to develop plans to safely evacuate disabled visitors. The illuminated surface of the exit sign must be rated not less than 5 footcandles to meet OSHA requirements of 29 Code of Federal Regulation (CFR) 1910.37. NFPA 101 also requires 5 footcandles for internally and externally illuminated signs, with some exceptions, such as approved self-luminous or electroluminescent signs that provide evenly illuminated letters. Signs must not contain decorations, furnishings, or pieces of equipment that impair visibility of an exit sign.

Never place other brightly illuminated signs, displays, or objects in the line of vision of any required exit sign. Externally illuminated exit signs use a light source located outside of the device or legend to be illuminated. Internally illuminated exit signs use a light source located inside the device, such as incandescent, fluorescent, electroluminescent, light-emitting diodes, or self-luminous lighting. Self-luminous exit signs use a light source illuminated by self-contained power sources such as tritium and operate independently of external power sources, with the light source contained inside the device. Electroluminescent exit signs use a light-emitting capacitor with the light source contained inside the device.

Emergency Lighting

NFPA 101 requires emergency illumination to provide a minimum of 1.5 hours of light. Arrange lighting to provide initial illumination of not less than an average of 1 footcandle. This level can decline to a minimum of 0.6 footcandle average at any point at the end the lighting time of 1.5 hours. Candle units measure the intensity of visible light, whereas a lumen refers to the rate of flow of light or luminous flux. A single lumen produces a flux on 1 ft^2 of a sphere one foot in radius, with a light source of one candle at the center. It radiates uniformly in all directions. One "lux" equals a unit of illumination equal to one lumen per square meter. A footcandle directly measures visible radiation falling on a surface. A footlambert unit measures the physical brightness on any surface emitting or reflecting visible light.

OSHA Egress and Exit Standards

OSHA Standard 29 CFR 1910.35 defines a means of egress as any continuous and unobstructed way of exit travel from any point in a building or structure to a public way. Egress consists of three separate and distinct parts:

- Exit access—portion that leads to the entrance of an exit
- Exit—portion separated from all other spaces of a building or structure by construction or equipment to provide a protected way of travel to the exit discharge
- Exit discharge—portion between the termination of an exit and a public way

OSHA Standard 29 CFR 1910.37 requires marking of exits with a visible sign. Mark exit access routes not immediately visible to occupants by using readily visible signs. Provide a sign that reads "Not an Exit" for any door, passageway, or stairway that occupants could mistake for an exit or exit access. OSHA permits signs that indicate actual nonexit locations, such as "To Basement or Storeroom." Signs designating an exit or a way of exit access must be distinctive in color and provide a contrast with decorations, interior finish, or other signs. Every sign must have the word *Exit* in plainly legible letters not less than 6 in. high, with the principal stroke of the letter 0.75 in. wide. Use arrows to direct occupants to the proper exit direction when confusion could arise.

Portable Fire Extinguishers (29CFR 1910.157)

The OSHA standard addresses the requirements for portable fire-extinguishers in general industry. OSHA requires the use of extinguishers approved by a recognized testing laboratory such as Factory Mutual (FM) Global or listed by Underwriters Laboratories (UL). OSHA requires extinguishers of the proper type, size, and quantity for the class of fire expected. Install extinguishers in locations with easy accessibility to users. Inspect and maintain extinguishers to ensure they remain in good operating condition. Ensure that only trained and designated persons use the extinguishers. Where the employee has provided fire extinguishers for employee use, the employer shall provide an educational program to familiarize employees on the principles and use of the extinguishers. This educational program should be completed during the initial hiring and annually thereafter. Annual retraining must address their operation and safe use.

Maintenance

Maintenance requirements can vary, depending on the type of portable fire extinguisher. Stored-pressure or dry-chemical type extinguishers do not require an internal examination. Water or steam types of fire extinguishers require discharge, disassembly, and inspection annually. Dry-chemical extinguishers undergoing a 12-year hydrostatic test will require emptying and are subjected to applicable maintenance procedures every six years. OSHA exempts nonrefillable, disposable, dry-chemical extinguishers from this requirement. Follow requirements found in the manufacturer's suggested maintenance procedure as presented in the operating manual. Hydrostatic testing of portable fire extinguishers can protect against unexpected service failure because of internal corrosion, external corrosion, or damage from abuse. Perform hydrostatic testing using trained personnel with proper equipment and facilities. Table 1 of 29 CFR 1910.157 provides test intervals for various types of extinguishers (Table 7.2).

Table 7.2 OSHA Fire Extinguisher Requirements

- Fire extinguishers located in assigned places
- Fire extinguishers access not blocked or hidden
- Fire extinguishers mounted to meet NFPA 10 requirements
- Pressure gauges show adequate pressure
- Pin and seals in proper place
- Fire extinguishers show no visual sign of damage or abuse
- Nozzles free of blockage

Portable Fire Extinguishers

Potable extinguishers apply an agent developed to cool burning fuel, displace or remove oxygen, or stop the chemical reactions that suppress or extinguish the fire. Compressing the handle of an extinguisher opens an inner canister of high-pressure gas that forces the extinguishing agent from the main cylinder through a siphon tube and out the nozzle. When using a portable fire extinguisher, use the PASS Model. The four steps include *pulling* the pin on the extinguisher, *aiming* the nozzle at the base of the fire, *squeezing* the handle firmly, and *spraying* using a sweeping motion.

Use Class A extinguishers for fires involving combustibles such as wood, paper, some plastics, and textiles. This class of fire requires the heat-absorbing effects of water or the coating effects of certain dry chemicals. Extinguishers appropriate for Class A fires should contain a green triangle with the letter *A*. The travel distance for Class A extinguishers must not exceed 75 ft.

Use Class B extinguishers for fires involving flammable solvents and gasoline. Class B fire extinguishers cut off oxygen supply to the fire, which hinders release of combustible vapors. Extinguishers appropriate for Class B fires should contain a red square with the letter *B*. The travel distance for Class B extinguishers must not exceed 50 ft.

Use Class C fire extinguishers for fires involving live electrical equipment or circuits. These types of fires require the use of nonconductive extinguishing agents. You can use Class A or B extinguishers on de-energized electric equipment. Extinguishers appropriate for Class C fires should contain a blue circle with the letter *C*. Base the placement of Class C extinguishers on the appropriate A or B hazard present in the area.

Use Class D fire extinguishers for fires involving combustible metals such as magnesium, titanium, potassium, calcium, sodium, and lithium. Most metal fires come from metal shavings or dust from machining work. Applying water on a Class D fire can help energize the flames and creates additional heat. Recommend using dry powder extinguishing agents on Class D fires. Extinguishers appropriate for Class D fires should contain a yellow five-point star containing the letter *D*. The travel distance for Class D extinguishers should not exceed 75 ft.

Use Class K fire extinguishers for fires involving cooking equipment that use fats, grease, or oils. These types of fire extinguishers operate on the saponification principle. Saponification occurs during the application of alkaline mixtures, such as potassium acetate, potassium citrate, or potassium carbonate, to burn cooking oil or fat. The alkaline mixtures combined with the fatty acid create on the surface a soapy foam that hinders vapors and steam, resulting in extinguishing the fire. Extinguishers appropriate for these fires contain the letter *K*.

Marking Extinguishers

NFPA 10 provides recommendations for marking extinguishers and/or fire extinguisher locations. Extinguishers suitable for more than one class of fire may be identified by multiple symbols. Apply markings to the extinguisher on the front with large letters to ensure easy reading at a distance of 3 ft. Extinguishers must contain information on the label about water equivalency rating and the amount of square footage that the extinguisher can handle if operated by a trained individual. An extinguisher marked 4A contains the equivalent of 5 gal of water. B/C ratings indicate square footage effective for areas up to 25 ft^2. No ratings are established for Class C or D fires. OSHA requires selection and placement of fire extinguishers based on the class and size of fires anticipated in the covered area. Install an adequate number of portable extinguishers, appropriate for the types of areas being protected, throughout a facility. Inspect extinguishers monthly, and maintain them according to manufacturer's requirements. Use only fire extinguishers approved by a testing organization such as FM or UL.

Flammable and Combustible Materials

Flammable and combustible liquids, vapors, and gases present major fire hazards. Train workers to recognize these types of hazards. The vapors of flammable liquids can ignite in a number of ways, such as a spark from a motor, friction, or static electricity. A flash point is the temperature at which a liquid gives off enough vapor to form an ignitable mixture when mixed with air. When a liquid reaches its flash point, contact with any source of ignition will cause the vapor to burst into flame. Use approved containers and portable tanks when handling and storing flammable and combustible liquids. Provide an adequate number of portable fire extinguishers near areas with flammable and combustible materials. Post warning signs to inform people of the hazards present in these systems. A combustible liquid has a flash point at or above 100°F (37.8°C). A flammable liquid has a flash point below 100°F (37.8°C).

Flammable or combustible liquid storage and usage containers must meet requirements contained in NFPA 30, Flammable and Combustible Liquids Code. Store combustible waste material such as oily shop rags and paint rags in covered metal containers. Post all storage areas as "NO SMOKING" as required by 29 CFR 1910.106, the OSHA Flammable and Combustible Liquids Standard. Piping systems including tubing, flanges, bolting, gaskets, valves, fittings, and all pressurized parts containing flammable and combustible liquids must meet the requirements of NFPA 30. Label piping in accordance with ANSI/ASME standards. Post each inside-storage location as a "NO SMOKING" area. Consider NFPA 30, *The Flammable and Combustible Liquids Code Handbook*, as an essential reference for those involved in the distribution or use of flammable and combustible liquids. The code covers general provisions, tank storage, piping systems, container and portable tank storage, and operational requirements. The code contains information about safety-can requirements and the design, construction, and testing of safety cabinets.

Fire and Chemical Hazard Symbols

NFPA 704, Standard System for Identification of the Hazards of Materials for Emergency Response, helps identify dangerous characteristics of hazardous materials. NFPA 704 applies to industrial, commercial settings, and institutional facilities that manufacture, process, use, or store hazardous materials. NFPA 704 does not apply to transportation requirements, general public use, or occupational exposures. NFPA 704–mandated labels enable emergency responders to make quick decisions about evacuation and other emergency control operations. The NFPA 704 Diamond uses the colors of blue, yellow, red, and white. Blue indicates a health hazard. Yellow indicates an instability or reactive hazard. Red indicates a fire hazard. White indicates "Special Notice" and can contain symbols indicating the presence of an oxidizer, an asphyxiate, or a waterreactive hazard. (Note: The number 0 indicates a minimal hazard, whereas 4 indicates the most severe hazard.)

Electrical Equipment Installations

Electrical equipment should be maintained and installed to meet NFPA/ANSI 70, National Electrical Code (NEC)®. Use only UL-listed or FM-approved equipment in areas with flammable gases or vapors. Temporary or makeshift wiring, particularly if defective or overloaded, can result in electrical fires. Overloaded or partially grounded wiring may also heat up enough to ignite combustibles without blowing fuses or tripping circuit breakers. When flammable liquids are used or stored, provide bonding and grounding with adequate and true grounds in accord with the NEC.

OSHA Emergency Action Requirements

OSHA requires all covered entities to develop an emergency action plan. Employers can refer to 29 CFR 1910.38 for the OSHA Emergency Action Plan requirements. Keep a written plan on file at the work location, and make the document available to employees for review. However, an employer with 10 or fewer employees may communicate the plan orally to employees. An emergency action plan must include at a minimum procedures for reporting a fire or other emergency. The plan must also detail types of evacuations and exit route assignments, including procedures to follow for those who remain to operate critical operations before evacuating. The plan must cover procedures or processes for accounting for all employees after evacuation. OSHA requires direction and guidance for employees performing rescue or medical duties. Employers must provide names or job titles of individuals who can provide additional assistance to employees seeking additional information about the plan or an explanation of their duties under the plan. Employers must maintain an employee alarm system. The employee alarm system must use a distinctive signal for each purpose and comply with OSHA requirements published in 29 CFR 1910.165. Senior management must designate and train employees to provide assistance to others and ensure safe and orderly evacuations. An employer must review the emergency action plan requirements with each employee on plan development or initial job assignment. Employers must also review the plan when an employee's responsibilities under the plan change or whenever the plan undergoes revision.

Emergency Management

Emergency managers and coordinators must consider and take into account all hazards, all phases, all impacts, and all stakeholders relevant to emergencies or disasters. We can define emergency management as "the process related to preparing, mitigating, responding, and recovering from emergency or disaster events." Emergency planning must address all hazards, phases, impacts, and stakeholders. Mitigation consists of those activities designed to prevent or reduce losses from disaster. It is usually considered the initial phase of emergency management, although it may be a component of other phases. Preparedness focuses on the development of plans and procedures to guide effective response efforts. The response phase concerns the initial and immediate reaction to the situation. Recovery consists of activities that continue beyond the initial emergency period and focus on issues such as critical community capabilities and organizational functions. Recovery can extend into the period of reconstruction.

The phases of emergency preparedness often overlap. However, effective emergency management coordinates the activities in all four phases. Emergencies and disasters can cut across a broad spectrum and can include infrastructure, human services, and the economy. When considering the "all hazards" approach, planners must also consider any predictable consequences relating to those hazards. All stakeholders must plan to focus on the management principles of coordination and collaboration.

Effective emergency management requires close working relationships at all levels, including government, private sectors, and the general public. Emergency managers and planners must give a greater attention to the importance of prevention and mitigation activities. Emergency managers must learn to think strategically and understand the important role of serving as an advisor. Strategic management must focus on issues such land-usage planning, environmental management, building-code enforcement, and training with realistic exercises.

Organizational Emergency Medical Care and First Aid

OSHA requires covered entities to plan for emergency first aid and medical treatment. Each facility must plan for providing emergency medical care to employees and others present on the site. Employers must provide for competent emergency medical care on all shifts. The emergency action or operation plan must outline standardized protocols. Ensure all on-site emergency medical personnel achieve certification on basic first aid and cardiopulmonary resuscitation levels. Employers must coordinate the use of appropriate off-site emergency medical care providers when necessary. Emergency plans must contain procedures for directing outside emergency medical professionals to the correct location. Employers must also provide first aid supplies appropriate for the workplace. Organizational members must understand how to use and implement the emergency medical plan. Communicate emergency information to outside contractors, vendors, and visitors.

Dealing with Risk

Emergency managers must endeavor to use sound risk-management principles, such as hazard identification and analysis, when assigning priorities and allocating resources. NFPA 1600 states that emergency management programs "shall identify hazards, monitor those hazards, the likelihood of their occurrence, and the vulnerability of people, property, the environment, and the entity [program] itself to those hazards." The Emergency Management Accreditation Program (EMAP) Standard echoes this requirement for public sector emergency management planning. Effective risk management helps identify natural and man-made hazards that could impact the community or organization. Hazard analysis must consider vulnerabilities and their impact on specific communities or organizations. Mitigation strategies, EOPs, and continuity or recovery plans address specific risks. Other important considerations include budgets, human-resource decisions, public education sessions, and training/exercises. Emergency managers must work to achieve a unity of effort among all sectors. Focus on ensuring the integration of all emergency plans. Integration of the plans must support the community's vision and values.

Private sector continuity plans should take into account the community emergency operational planning. Businesses can provide significant resources during disasters and can serve as a critical component of the community's EOP. Local emergency management plans must also synchronize with higher-level plans and procedures. Creating a team atmosphere can help build a consensus and facilitate communication. Planners must understand the difference between "collaboration" and "coordination." Coordination ensures the identification and accomplishment of functions, roles, and responsibilities. View collaboration as the attitude, climate, or culture that characterizes the cooperation that exists within a community. Collaboration creates the environment in which coordination can function effectively. The principle of coordination requires emergency planners to achieve agreement of common purpose among a variety of entities. This agreement helps the variety of independent activities accomplished by these entities to contribute to the common purpose. Flexibility remains the key to successful emergency management. Emergency management requires the use of science and knowledge-based approaches. Professionalism in the context of emergency management pertains to the concept of commitment. Professional emergency managers should study past disasters and try to understand social issues related to emergency or disaster events. Professionals must possess the knowledge and understanding of emergency management practices, standards, and guidelines.

Disaster Management and Business Continuity

NFPA 1600 focuses on planning for disasters and emergency events. The standard also established a common set of criteria for disaster management, emergency management, and business continuity programs. NFPA 1600 identifies methodologies for exercising emergency plans. It provides a listing of resource organizations within the fields of disaster recovery, emergency management, and business continuity planning. The standard assists organizations with their planning in areas of mitigation, preparation, response, and recovery. Organizations must establish an emergency management planning committee. Conduct a hazard vulnerability analysis (HVA) to determine events and hazards likely to occur in the facility and community. Ensure the plan addresses mass casualty situations, including terrorists' events of a chemical, biological, or radiological nature. The plan should address risk management principles and provide for a command structure to assess situations, coordinate actions, and make decisions. A good HVA will identify and rate potential emergencies or disaster situations that could occur in the facility or community.

Establish priorities, and rank each emergency situation identified by the HVA. Determine the organization's role in the community plan. Use an analysis chart to guide the process and help with assignment of probabilities. Some experts recommend use of a simple numerical system to help assess and prioritize each potential risk. Evaluate the HVA at least annually, and more frequently, if warranted because of location or risk. Provide realistic training and education for all emergency personnel. Ensure everyone understands their roles and responsibilities. Validate individual understanding during emergency exercises or drills. Educational sessions can help reduce fear among personnel responding to terrorism-type events. All hazard planning is based on an HVA. The EOP 51 describes the collective procedures and processes used for all hazard response and the early recovery phase. Emergency operations planning should address the potential surge and organizational ability to maintain normal operations. The most successful organizations do constant self-evaluation. Organize EOPs around functions and not particular hazards. Response-generated demands include things such as good decision making, reliable communication, and interagency coordination.

Emergency Management Planning

Consider ways to improve management involvement. Determine if drills address actual risks identified in the hazard analysis. Ensure the plan reflects on lessons learned from drills and actual events. Ensure the plan reflects the physical layout of the facility or process. Ensure that emergency training meets established objectives. The plan must contain current names, agencies, and contact numbers as required for response. Ensure community agencies receive updates on plan changes. Modify plans after each drill, exercise, or actual emergency. Update plans when personnel or their responsibilities change or the design of the facility changes. Ensure each drill or exercise is evaluated by a responsible person to determine performance and opportunities for improvement. Critique drills to identify strengths and weaknesses. Focus on execution issues, training deficiencies, and problem areas.

Incident Command System

The Incident Command System (ICS) organization relies on the functions command, operations, planning, logistics, and finance/administration to manage emergency or disaster incidents. The system coordinates responses for events involving multiple jurisdictions or agencies. It retains the principle of a unified command for coordinating the efforts of many jurisdictions. The system must ensure joint decisions in the areas of objectives, strategies, plans, priorities, and public communications. The system focuses on responder readiness to manage and direct incident actions by coordinating issues before an event. The benefits of using ICS can include maintaining a predictable chain of management accountability and providing flexible response to specific emergencies. ICS can also improve documentation procedures, provide a common language to facilitate outside assistance, and prioritize response checklists or protocols. Recognizing that many activities must occur to manage an emergency incident remains a primary tenet of ICS. We can group these tasks into categories that reflect functional similarities. The command function provides overall direction of the response through the establishment of objectives for the organization to meet. Consider other important management issues:

- Safety: identify and assess hazards to prevent injury or illness.
- Liaison: Provide coordination/integration with agencies or organizations external to the response system.
- Public information: Develop/provide incident information for both the public and response personnel.
- Senior advisors: additional positions, as designated by the incident commander, to provide needed advice and expertise to the command staff.
- Operations: through developed strategies and tactics, achieve the command objectives.

Operations and Planning

The command or operations function sets the goals and objectives of the response for the organization, and the operations function performs activities to achieve them.

The plans function supports the response organization by conducting the incident-planning activities and by acquiring, processing, documenting, and disseminating all incident-related information.

Logistics

This function supports the response organization with facilities, transportation, supplies, equipment maintenance and fuel, food services, communications and information technology (IT) support, and emergency responder medical services.

Finance or Administration

This function supports the response organization by tracking incident costs and addressing issues such as reimbursements, claims, and regulatory compliance.

Communication and Information Management

Responders and managers in all agencies and jurisdictions must have a common operating picture for a more efficient and effective incident response.

Common Incident Command System Principles and Terms

Consider the following benefits when employing an ICS.

The use of common terminology for resource descriptions, organizational functions, and incident facilities across disciplines can contribute to improve effectiveness. Integrated communications

Table 7.3   Emergency Assessment Topics

- Hazard type, impact, and expected duration
- Secondary hazards and hazard impacts
- Weather and other geophysical conditions, including warning periods and duration
- Impact of timing and duration, such as day and time of onset
- Response actions, locations, and scope of impact
- Predicted expansion and/or migration of impact area
- Specific needs, including type, amount, and location
- Priority of resource needs

help send and receive information within the organization and externally to other disciplines. The modular organization structure permits assets within each functional unit to expand or contract based on the requirements of the event. The unified command structure allows response organizations to work through designated managers to establish common objectives and strategies, which can reduce conflict and duplication. The span-of-control principle ensures that each supervisory level oversees an appropriate number of assets, based on size and complexity of the event. Consolidated incident action plans can effectively address all goals, objectives, strategies, and major assignments defined by the incident commander or by a unified command. Comprehensive resource management processes help to describe, maintain, identify, request, and track all resources within the system during an incident. Predesignating incident facilities can help with assigning locations where expected critical incident-related functions could occur. Assign duties to certain positions and never to specific individuals. The incident commander must maintain emergency command center effectiveness, ensuring communications and maintaining security. Key duties include providing public information and media releases, coordinating facilities, and sheltering, feeding, and counseling as needed. Using the concept known as management by objective during incident action planning can help achieve objectives when addressing multiple considerations.

Situational awareness refers to a person's knowledge of the situation as related to the evolving state of the event environment. Situation assessment during emergency response and recovery combines information about incident geography, topography, weather, hazard or hazard impact, and resource data. This assessment provides a foundation for decision making (Table 7.3).

Standardization

Standardized processes and methods involve well-described, reproducible, and usually sequential steps to accomplish a stated objective. The EOP guides response and recovery using standardized formats and the ICS. Standardized processes contribute to efficiency in all four phases of emergency management. Use standardized templates from the ICS to help develop incident action plans, conduct briefings, and complete situation reports.

Community Involvement

Federal Emergency Management Agency (FEMA) Publication H-34, entitled *Are You Ready: A Guide to Citizen Preparedness*, contains facts on disaster survival techniques, disaster-specific information, and how to prepare for and respond to both natural and man-made disasters.

Health-care organizations must adopt a community-wide perspective when planning for mass casualty incidents. Senior leaders must maintain good relationships with response agencies in the community, including other area health-care facilities. Clinics and nursing homes may play key roles in large disasters. Public health departments will usually institute appropriate public health interventions, including immunizations and prophylactic antibiotics. Establish working relationships with all responders, including local emergency management agencies, law enforcement personnel, and local fire officials.

Health-Care Facilities

Health-care organizations must work to help assess community health needs and available resources to treat evacuees from other areas. The organization must determine community priorities identified in the HVA. Clarify the organization's role during the annual community-wide emergency exercise. Develop plans for coordinating with the media. Establish a media briefing area, with established security procedures. Establish procedures for ensuring the accuracy and completeness of all information approved for public release. Determine an appropriate method of communicating technical or medical information.

Health-care facilities near transportation systems should maintain a 2-day supply of drugs and medical supplies. Remote and rural facilities should plan for maintaining operations for least 96 hours. Remote facilities should plan to obtain drugs and other supplies from existing vendors, group purchasing organizations, other hospital systems, and state hospital associations. The Centers for Disease Control and Prevention (CDC) National Pharmaceutical Stockpile will normally make drugs available as soon as possible after detection of a biological or chemical agent. The stockpile serves as a national repository of antibiotics, chemical antidotes, antitoxins, life-support medications, intravenous-administration supplies, airway-maintenance supplies, and medical/surgical items. Consider the stockpile as a supplemental source of supplies and not as a first-response source. Hospitals provide decontamination and treatment for any and all victims. Health-care facilities many times can provide hazardous waste operations and emergency response training for local governments and businesses. They can also provide information and services related to emergency planning. Hospitals now actively participate in community-based planning and preparedness activities.

Weather-Related Emergencies

Tropical Storms and Hurricanes

The National Weather Service (NWS) issues warnings when tropical storms, hurricanes, or tropical storms threaten the United States and its territories. As soon as conditions intensify to the tropical storm level, the storm receives a name, and the NWS begins issuing advisories and watches. The NWS issues a tropical storm watch when any system exhibiting sustained winds of 39–73 mph could hit a coastal area within 48 hours. The NWS upgrades the watch to a warning when the system could hit within 36 hours. When a tropical system exhibits sustained winds of 74 mph or higher, the NWS issues hurricane watches or warnings. Authorities issue a hurricane watch 48 hours in advance of the anticipated onset of tropical-storm-force winds. The watch can become a hurricane warning 36 hours in advance of the anticipated onset of tropical-storm-force winds.

Thunderstorms

Previously, the NWS issued severe thunderstorm warnings whenever a thunderstorm is forecast to produce wind gusts to 58 mph (50 knots) or greater and/or hail size of 3/4 in. (penny-size) diameter or larger. For several years, offices that cover areas of Kansas experimented using a warning criterion of 1 in. diameter hail. During the spring and early summer of 2009, this experiment expanded to other areas in the central and western United States. Beginning January 5, 2010, the minimum size for severe hail nationwide increased to 1-in. (quarter-size) diameter. There will not be a change to the wind gust criterion of 58 mph. This change is based on research that indicated significant damage does not occur until hail size reaches 1-in. (quarter-size) in diameter.

Floods

Floods remain the most common and widespread of all natural disasters. Most communities in the United States can experience some degree of flooding after spring rains, heavy thunderstorms, or winter snow thaws. Most floods develop slowly over a period of days. Flash floods, however, are like walls of water that develop in a matter of minutes. Flash floods can be caused by intense storms or dam failure. Tune to NOAA Radio and be prepared to evacuate. Tune to local radio and television stations for additional information.

Tornadoes

Tornado wind speeds can exceed 300 mph. Their width varies from 50 yards to 1 mile or more. They can travel 5–50 miles or more moving at 30–75 mph. Tornadoes sometimes reverse or move in circles. The NWS issues tornado watches to alert the public that conditions are favorable for the development of tornadoes in and close to the watch area. Local NWS offices can issue warnings to the public when trained spotters, law enforcement, or radar indicate a tornado in the area. The NWS issues warnings with information about the tornado's location and its anticipated path.

Winter Storms

Severe winter storms bring heavy snow, ice, strong winds, and freezing rain. Winter storms can prevent employees and customers from reaching the facility, leading to a temporary shutdown until roads are cleared. Heavy snow and ice can also cause structural damage and power outages. Winter storms vary in size and intensity. A winter storm watch indicates severe conditions may affect the area. The storm could produce freezing rain, sleet, or heavy snow. A winter storm warning indicates the approach of severe winter weather conditions. A heavy snow warning indicates the expectation of snowfall in large amounts. The actual amount can vary by geographic location. Blizzard warning indicates the possibility of sustained wind speeds of at least 35 mph that are accompanied by considerable falling and/or blowing snow.

Wildfires

Wildfires can pose great threats to life and property. Those at risk should develop procedures to address the risks to the facility's operation and individual safety. Many western states experience wildfires frequently, but other states also experience the risk of wildfire occurrence. Address these events when conducting the HVA. The National Interagency Fire Center (NIFC), located in

Boise, Idaho, serves as the country's support center for wildland firefighting. Seven federal agencies from the Departments of the Interior, Agriculture, and Commerce make up the contributors of the fire-center operation. The fire center website provides information on safety training, safety advisories, and fire incident reporting.

Earthquakes

Earthquakes can seriously damage buildings, take life, and destroy infrastructure. They can also disrupt gas, electric, and telephone services and trigger landslides, avalanches, flash floods, fires, and tsunamis. Aftershocks can occur for weeks following an earthquake. In many buildings, the greatest danger to people is when equipment and nonstructural elements such as ceilings, partitions, windows, and lighting fixtures become hazards. Earthquakes occur most frequently west of the Rocky Mountains but can happen in other locations such as the central Mississippi River valley.

Technology Emergencies

These emergencies include any interruption or loss of a utility service, power source, life-support system, information system, or equipment needed to keep the business in operation. Identify all critical operations, including electric power, gas, water, hydraulics, compressed air, municipal and internal sewer systems, and wastewater treatment services. Also consider security and alarm systems, elevators, lighting, heating, ventilation, air-conditioning systems, and electrical distribution systems. Evaluate transportation systems, including air, highways, railroads, and waterways. Determine the impact of service disruption.

Transportation Accidents

The Department of Transportation (DOT) regulates the movement of hazardous chemicals. When hazardous chemicals that would pose a significant hazard to the public if released from their packing are transported interstate, they must be labeled with appropriate words of identification and caution. Shipping papers identifying the hazardous material being transported are required to be in the vehicle or vessel. Major transportation accidents often cause chemical spills, fires, explosions, and other problems, which call for special operations such as rescue and evacuation.

Bomb Threats

Preparing a questionnaire with relevant questions placed near the phone can be very useful. The information gathered on this questionnaire may be sufficient to discount the threat, or may direct that actions other than evacuation to be taken. It is not as unlikely as you may think for the caller to give his/her address. If a suspicious object is located and thought to be a bomb, the Federal Bureau of Investigation (FBI) recommends evacuation immediately. Contact local law enforcement personnel or your Emergency Management Agency to secure the services of the nearest explosive/bomb disposal team for assistance and training. According to the FBI, many callers ask to speak with a specific person. The questionnaire may be the most important resource for dealing with bomb threats and documenting important information.

Information and Communication Emergencies

One of the primary challenges to responders is the chaos imposed on information. The ability to access pertinent information as needed is paramount to a successful response and recovery. The art of gathering information, analyzing and summarizing it, and then sharing it with those who need it is known as information management.

Identify key communications and IT components that are critical to the continuation of essential services in an emergency. Also specify any procedures to be followed in the hours preceding a storm to protect computers, paper records, securing equipment, placing garbage bags over files, or moving files upstairs. Identify which, if any, databases should be backed up at the last possible moment. Consider the service provider's infrastructure and facilities, and consider the use of alternate providers. Periodically test all redundant communications systems, and consider the use of divergent routes.

Telecommunications Service Priority Program

This program provides organizations engaged in national security and emergency preparedness functions with priority provisioning and restoration of telecommunications services that are vital to coordinating and responding to crises. A telecommunications service user with a Telecommunications Service Priority (TSP) assignment is assured of receiving service by the service vendor before a non-TSP service user.

Government Emergency Telecommunications Service Program

The Government Emergency Telecommunications Service (GETS) program provides emergency access and priority processing in the local and long-distance segments of the public switched network. It is to be used in an emergency or crisis situation during which the probability of completing a call over normal or other alternate telecommunication means has significantly decreased.

Wireless Priority Service

Wireless Priority Service (WPS) can improve connection capabilities for a limited number of authorized national security and emergency-preparedness cell phone users. In the event of congestion in the wireless network, an emergency call using WPS will have priority queuing for the next available channel. Learn the capabilities of the system so as to maximize the value of the plan. Obtain a last-resort backup means of communication such as wireless, Wi-Fi, or satellite. Consider high-frequency (HF) radio as an option, recognizing that HF usually requires a skilled operator such as a licensed HAM radio operator. Evaluate the resiliency, redundancy, and interoperability of the system while performing your inventory and risk assessment analysis. These steps should consider the following:

- Diversity of communications systems
- Facility hardening and alternate routing
- Internal building infrastructure and hardware backup
- Uninterrupted power supplies
- Availability of replacement parts

Cyber Attack Incident Response

The array of data sources in emergency management is staggering. Data from voice, text, video, sensors, databases, forms, satellites, telemetry, and eyewitness accounts all play a role in managing disasters. Add to this the variety of data at various stages of planning, and the volume and sources of data can become overwhelming. Of paramount importance to management of this information and avoiding sensory overload is the capability to fuse various data sources into a coherent view. Imagine a screen with summary information on current bed status, patient status, resource availability, and links to up-to-date response plans and guides. The ability to access pertinent data in a timely manner is fundamental to success. Consider the following issues:

■ Defining scope and impact of potential problems
■ Isolating affected systems
■ Restoring automated systems and services
■ Notifying affected end-user supervisors and providing guidance on systems use
■ Taking immediate actions during the operational period (0–2 hours)

Homeland Security Exercise and Evaluation Program

The program exercises allow homeland security and emergency management personnel, from first responders to senior officials, to train and practice prevention, protection, response, and recovery capabilities in a realistic but risk-free environment. Exercises are also a valuable tool for assessing and improving performance while showing community resolve to prepare for major incidents. Through exercises, the Department of Homeland Security aims to help entities obtain objective assessments of their capabilities so that gaps, deficiencies, and vulnerabilities are identified and remedied before a real incident.

Planning for Terrorism

Educate and train staff about possible events and response actions. Experts advise local communities to be prepared to deal with the consequences of a terrorist event for the 12–36 hours before federal agencies can augment local response and provide specialized support. Potential risks associated with nuclear, chemical, biological, or radiological weapons by terrorists calls for sound emergency planning procedures. Terrorist events can result in potentially large numbers of casualties. The psychological impact of weapons of mass destruction and the relative ease of their acquisition poses a great threat.

In 2004, CDC published the *National Public Health Strategy for Terrorism Preparedness and Response Guide* to provide information on the following areas:

■ Detection, investigation, and laboratory sciences
■ Prevention programs, worker safety, and communication
■ Emergency response, research, and long-term consequence management
■ Workforce development

Biological Agents

Clinical symptoms may not appear for some time after an exposure. Biological agent organisms can live in a form of liquid droplets, aerosols, or dry. The signs and symptoms are usually nonspecific and may mimic natural infections such as the flu. Pathogens or disease-causing organisms include bacteria, viruses, and fungi. Viruses are submicroscopic organisms that require living cells to reproduce and multiply. Be aware of signs and symptoms that develop at an uncharacteristic time of the year, in an unusual pattern, or in a normally healthy population. Be alert for groups of patients from a single location or event. Watch for lower incidences of symptoms among people who have been indoors.

Bacteria, Viruses, and Toxins

Bacteria are self-sustaining organisms that do not require a host to reproduce. Examples can include anthrax, plague, and cholera. Viruses are much smaller than bacteria and need a host to survive. The host can be plants, animals, insects, bacteria, or humans. Examples are smallpox, Venezuelan equine encephalitis (VEE), and Ebola. Living organisms can produce toxins or poisonous chemical compounds. These agents show lethality about 1000 times higher than standard chemical agents. Toxins normally do not pose an absorption risk. Toxins include botulism and ricin.

Chemical Agents

Classify these agents into general categories of blood, blister, choking, irritating, and nerve classifications. The toxicity, mode of action, and effects can vary, depending on the agent. Chemical agents act within minutes, and people exposed will develop symptoms immediately. Consider inhalation as the primary route of exposure. Toxicity of an agent depends on the size of the particles and the water solubility of the gas. Small particles and gases with low solubility enter more deeply into the lungs. Quick decontamination and immediate administration of an antidote are the best response to some agents. The following is a partial list of chemical emergency response cards and is available on the CDC website:

- Hydrogen cyanide (AC)
- Lewisite (L, L-1, L-2, L-3)
- Mustard gas (H)
- Nitrogen mustard (HN-1, HN-2, HN-3)
- Potassium cyanide (KCN)
- Sulfur mustard (H)

Blood Agents

These agents interfere with the ability of the blood to transport oxygen. Consider all blood agents as toxic at high concentrations. Exposure can lead to rapid asphyxiation and death. Symptoms can include respiratory distress, vomiting, diarrhea, vertigo, and headaches. Fresh air and respiratory therapy may help some victims. An example of a blood agent is cyanide.

Blister Agents (Vesicants)

These agents cause burns to the eyes, skin, and respiratory tract tissues. They can penetrate clothing and be absorbed into the skin. Symptoms can include tearing of the eyes, swollen eyelids, itching, burning pain, and blisters in warm, moist areas of the body. Watch for a burning sensation in the nose and throat, hoarse voice, shortness of breath, cough, abdominal pain, and diarrhea. Mustard is a common type of blister agent.

Choking Agents

These agents stress the respiratory tract and can result in asphyxiation. An edema can develop in the lungs, and patient symptoms may resemble those of a drowning victim. Symptoms include eye irritation, choking and coughing, and respiratory distress. Victims may smell like chlorine or newly cut hay (phosgene). An example of a choking agent is phosgene.

Irritating Agents

These agents cause respiratory distress and tearing with the intention of incapacitating the victim. Symptoms include severe pain to the skin, burning and irritation of the eyes and throat, respiratory distress, coughing, choking, nausea, and vomiting. Most exposed people smell of pepper or tear gas. Examples of irritating agents include tear gas and pepper spray.

Nerve Agents

These agents are the most toxic chemical agents and can cause death in minutes. They can be inhaled or absorbed through the skin. Nerve agents can affect organs with muscles and glands. Exposure results in increased secretions of saliva, tears, and nasal mucus. Other symptoms include secretions in airways and gastrointestinal tracts, along with sweating, muscle contractions, and hyperactivity of the digestive tract. Some victims show symptoms of twitching, weakness, and hypertension. Sarin is an example.

Industrial Chemical Agents

There are a wide variety of potential chemicals, including organic-based phosphate pesticides such as Malathion, that could be used for malicious purposes. These compounds disrupt the acetyl cholinesterase enzyme just as nerve agents do. For example, arsenic trioxide acts as a metallic poison.

Nuclear Devices

A nuclear terrorist incident can involve the detonation or threatened detonation of a nuclear bomb or the detonation or threatened detonation of an explosive device that includes nuclear materials. Terrorists could also cause a nuclear incident by detonating an explosive device near a nuclear power plant or attacking nuclear cargo during transport. Terrorists could contaminate food or other products with radioactive materials. Simple radiological devices such as an isotope, if placed in public, could spread radiation without the use of an explosive device.

Nuclear Incident Response Team of Department of Energy

This team provides expert personnel and specialized equipment to a number of federal emergency-response entities that deal with nuclear emergencies, nuclear accidents, and nuclear terrorism. Our emergency response personnel are experts in such fields as device assessment, device disablements, intelligence analysis, credibility assessment, and health physics.

National Nuclear Security Administration

The Office of the National Nuclear Security Administration provides support to the administrator and includes the functions of legislative affairs, public affairs, and liaison with other federal agencies; state, tribal, and local governments; and the public. It also provides support for resource management in the areas of budget formulation, guidance, and execution, personnel, and procurement management and the administration of contracts, as well as other activities as determined by the administrator.

National Institute of Occupational Safety and Health Publication Number 2002-139

This document identifies actions that a building owner or manager can implement without undue delay to enhance occupant protection from an airborne chemical, biological, or radiological attack.

National Disaster Medical System

The system interacts with the private enterprise and civilian volunteers to ensure availability of resources to support medical services following a disaster that overwhelms the local health-care resources. The overall purpose of the system is to establish an integrated national medical response capability for assisting state and local authorities in dealing with the medical and health effects of major peacetime disasters.

Pandemic Planning

FEMA and OSHA provide guidance with recommendations for developing plans to respond to a pandemic. The focus is on planning during the interpandemic period for issues such as surveillance, decision-making structures, communications, education and training, patient triage, clinical evaluations, admission, facility access, occupational health, distribution of vaccines, antiviral drugs, surge capacity, and mortuary issues. The activities suggested in Supplement 3 are intended to be synergistic with those of other pandemic influenza planning efforts, including state preparedness plans. US health-care facilities must be prepared for the rapid pace and dynamic characteristics of pandemic influenza. All hospitals should be equipped and ready to care for a limited number of patients infected with a pandemic influenza virus or other novel strains of influenza.

Review Exercises

1. Describe the major elements of an OSHA-mandated fire prevention plan.
2. Describe the four stages of fire development.
3. What is the purpose of NFPA 101?
4. List and describe the three levels of fire safety deficiencies.
5. List the four scoring areas of the Fire Safety Evaluation System.
6. List five priority emphasis areas of a quarterly fire inspection.
7. Define the concept known as "fire confinement."
8. Define the following terms:
   a. Exit access
   b. Exit
   c. Exit discharge
9. What is the flash point that determines whether a substance is flammable or combustible?

10. Explain the purpose of NFPA 704.
11. Describe the basic elements required in an OSHA-mandated emergency action plan.
12. What are the four phases of emergency management?
13. What is the purpose of NFPA 1600?
14. Define or describe the following terms:
    a. Integrated communications
    b. Modular organization structure
    c. Span of control
    d. Consolidated incident action plans
    e. Comprehensive resource management
    f. Management by objective
    g. Situational awareness
    h. Situation assessment
15. List five key emergency assessment topics.
16. What is the CDC National Pharmaceutical Stockpile?
17. Describe the parameters for the issuance of a severe thunderstorm warning.
18. What is the NIFC, and where is it located?
19. Describe the purpose of the following programs or services:
    a. The TSP Program
    b. The GETS Program
    c. The WPS
20. List the key issues to consider following a cyber attack.
21. What is the purpose of HSEEP?
23. Define or describe the following potential terrorist agents:
    a. Blood agents
    b. Blister agents (vesicants)
    c. Choking agents
    d. Irritating agents
    f. Nerve agents
    g. Industrial chemical agents

Appendix A: Hazard Control Management Evaluation Scoring System

Management Leadership and Employee Participation

IA: Hazard Control Policies and Procedures

(4) Workforce accepts, can explain, and fully understands the hazard control policy and procedures.
(3) Majority of personnel can explain policy and procedures.
(2) Some personnel can explain policy policies and procedures.
(1) Written (or oral, where appropriate) policies and procedures exist.
(0) No such policies or procedures exist.

Score: Comments:

IB: Goals and Objectives

(4) Workforce is involved in objectives development and can explain desired results/measures.
(3) Majority of personnel can explain desired results and measures for achieving them.
(2) Some personnel can explain desired results and measures for achieving them.
(1) Written (or oral, where appropriate) goals and objectives exist.
(0) No safety and health goals and objectives exist.

Score: Comments:

IC1: Leadership

(4) All personnel acknowledge that top management provides essential hazard control leadership.
(3) Majority of personnel see top management as active in hazard control function.
(2) Top management influence seen through safety presentations, training, and documents.
(1) Some evidence of top management involvement in hazard control efforts.
(0) No evidence of top management involvement in hazard control activities.

Score: Comments:

IC2: Leadership Example

(4) All personnel acknowledge that top management always sets positive example.

(3) Majority of personnel credit top management for setting positive examples.

(2) Top management generally sets positive example.

(1) Some evidence that top management generally says and does the right things to support hazard control.

(0) Top management does not appear to follow hazard control policies established for others.

Score: Comments:

ID: Organizational Member Involvement

(4) All personnel responsible for actively identifying and resolving hazard control–related issues.

(3) Most personnel believe their involvement helps identify and resolve hazard control–related issues.

(2) Some personnel believe their involvement positively impacts hazard control efforts.

(1) Employees generally believe that supervisors will consider their input and suggestions.

(0) Employee involvement in hazard control is not encouraged or rewarded.

Score: Comments:

IE: Hazard Control Responsibilities

(4) All personnel can explain organizational hazard control expectations and performance requirements.

(3) Majority of personnel can explain hazard control expectations and performance requirements.

(2) Some personnel can explain hazard control expectations and performance requirements.

(1) Performance expectations, including hazard control elements, spelled out for everyone.

(0) Specific hazard control responsibilities and performance expectations are generally unknown.

Score: Comments:

IF: Resources and Authority

(4) All personnel believe they posses necessary authority and resources to meet their responsibilities.

(3) Majority of personnel believe they possess necessary authority and resources to do their jobs.

(2) Authority and resources spelled out for all; some demonstrate a reluctance to use them.

(1) Authority and resources exist, but most still are controlled by supervisors.

(0) Authority and resources come from supervision without any delegation.

Score: Comments:

IG: Hazard Control Effectiveness

(4) Hazard control performance is measured against established goals, clearly displayed, and recognized.

(3) Personnel are held accountable for safe performance, with appropriate rewards and consequences.

(2) Accountability systems in place, but rewards/consequences do not always follow performance.

(1) Personnel generally held accountable, but consequences are more negative rather than positive.
(0) Accountability is inconsistent and mostly prompted by serious negative events.

Score: Comments:

IH: Hazard Control Quality Review

(4) In addition to a comprehensive review, a process is in place to drive continuous correction.
(3) Comprehensive review conducted annually and drives appropriate modifications.
(2) Quality review conducted but does not appear to drive all necessary changes.
(1) Changes policies driven by events such as accidents or compliance activities.
(0) No evidence of any quality or evaluation review process.

Score: Comments:

Hazard and Behavior Analysis

IIA1: Quality Review

(4) In addition to corrective action, regular expert surveys result in updated hazard inventories.
(3) Comprehensive expert surveys conducted periodically and drive appropriate corrective action.
(2) Comprehensive expert surveys conducted, updates, or corrective action sometimes lag.
(1) Qualified experts conduct surveys in response to accidents, complaints, or compliance activities.
(0) No evidence the organization conducts any comprehensive hazard surveys.

Score: Comments:

IIA2: Hazard Review

(4) All planned/new facility, process, material, or equipment fully reviewed by competent personnel.
(3) Hazard review of each planned/new facility, process, material, or equipment is conducted by experts.
(2) High-hazard planned/new facilities, processes, materials, and equipment are reviewed.
(1) Hazard reviews of new facilities, processes, materials, and equipment driven by problems exists.
(0) No system or requirement exists for hazard reviews of planned or new operations.

Score: Comments:

II A3: Job Hazard Analysis

(4) Employees are involved in the development of current hazard analysis on their jobs.
(3) Hazard analysis program exists for some jobs/processes and is understood by affected employees.
(2) Hazard analysis program exists for a few jobs/processes and is understood by affected employees.
(1) Hazard analysis program exists, but few employees are involved, and most are not aware of results.
(0) No routine hazard analysis process is in place at facility.

Score: Comments:

IIA4: Self-Inspection Program

(4) Employees/supervisors are well trained and conduct routine joint inspections with all findings corrected.

(3) Employees trained in inspection techniques and all routinely participate in workplace inspections.

(2) Routine inspections conducted by selected personnel, with appropriate corrective actions taken.

(1) An inspection program exists, but few employees are involved, and coverage/corrective actions are spotty.

(0) No routine inspection program is in place at facility.

Score: Comments:

IIB: Hazard Reporting and Correction

(4) Employees trained and empowered to correct any hazards identified by using own initiative.

(3) Comprehensive hazard data collection and analysis in place to drive the correction process.

(2) System exists for hazard reporting, and employees can use it, but system slow to respond.

(1) System exists for hazard reporting, but employees find it difficult to use or unresponsive.

(0) No hazard reporting system exists, or employees appear uncomfortable reporting hazards.

Score: Comments:

IIC: Accident Investigation and Root-Cause Analysis

(4) All loss-producing incidents and near misses investigated; root-cause analyses conducted as required.

(3) Occupational safety and health administration (OSHA) reportable incidents are investigated and effective preventions implemented.

(2) OSHA-reportable incidents investigated; cause determination and corrections are at times inadequate.

(1) Some investigations take place, but the root causes are seldom identified, and corrections are spotty.

(0) Incidents are not investigated, or investigations limited to reports required for compliance purposes.

Score: Comments:

IID: Accident Trending

(4) All employees are fully aware of incident trends, causes, and prevention methods.

(3) Trends are analyzed and displayed, common causes communicated, and management ensures prevention.

(2) Data are collected and analyzed, and common causes communicated to concerned supervisors.

(1) Data are collected and analyzed but not widely communicated for prevention purposes.

(0) No consistent effort to analyze incident data for trends, causes, and prevention.

Score: Comments:

Hazard Prevention and Control

IIIA: Controls

(4) Hazard controls concentrate on engineering fixes with reinforced/enforced safe work procedures.

(3) Controls are based on priority of engineering controls, work practices, administrative controls, and personal protective equipment (in that order).

(2) Hazard controls are fully in place, but the order of priorities varies with situation.

(1) Hazard controls are generally in place, but priorities and completeness vary.

(0) Hazard controls are incomplete, ineffective, or inappropriate in this workplace.

Score: Comments:

IIIB: Preventive Maintenance

(4) Personnel are trained to recognize and perform preventive, periodic, and routine maintenance as required.

(3) Effective preventive-maintenance schedule is in place and applicable to all equipment.

(2) Preventive maintenance schedule is in place and usually followed, except for higher priorities.

(1) A preventive maintenance schedule is in place but not performed as scheduled.

(0) Very little attention is paid to preventive-maintenance needs.

Score: Comments:

IIIC1: Emergency Planning

(4) All personnel know immediately how to respond as a result of effective planning, training, and drills.

(3) Most employees have understanding of responsibilities with regard to planning, training, and drills.

(2) Effective emergency response team is in place, but others are mostly uncertain of their responsibilities.

(1)Effective emergency response plan exists, but training/drills are inadequate and roles not well understood.

(0) Little effort is made to prepare for emergencies.

Score: Comments:

IIIC2: Emergency Planning (Equipment)

(4) Facility is fully equipped, with all systems/equipment in place/tested, and all personnel know what to do.

(3) Facility is fully equipped with appropriate emergency procedures, and most know what to do.

(2) Emergency procedures and equipment are in place, but only emergency teams know what to do.

(1) Emergency procedures and equipment are in place, but employees show little awareness.

(0) No real evidence exists of an effective effort at providing emergency equipment and other guidance.

Score: Comments:

IIID1: Medical Surveillance

(4) Occupational health providers are available onsite and involved in hazard identification and training.

(3) Occupational health providers are there as needed and are generally involved in assessment/ training.

(2) Occupational health providers are consulted about significant health concerns.

(1) Occupational health providers are available but normally concentrate on clinical issues.

(0) Occupational health provider assistance is rarely requested or provided.

Score: Comments:

IIID2: Injury Response

(4) Personnel fully trained in emergency medicine are always available.

(3) Personnel with basic first aid skills are always available, and emergency care is nearby.

(2) Personnel with basic first aid skills are usually available, and community assistance is nearby.

(1) Either onsite or nearby community aid is always available.

(0) Onsite or community aid cannot be ensured at all times.

Score: Comments:

Education and Training

IVA: Employee Training

(4) Employees are involved in hazard assessment and help develop/deliver training, and all are trained.

(3) Facility is committed to high-quality training, everyone participates, and regular updates are provided.

(2) Facility provides legally required training and makes an effort to include all personnel.

(1) Training is provided if needed, and experienced personnel are assumed to know the material.

(0) Facility depends on experience and informal peer training to meet needs.

Score: Comments:

IVB1: Supervisor Training

(4) All supervisors assist in worksite analysis, ensure physical protection, reinforce training, enforce discipline, and can explain work procedures.

(3) Most supervisors assist in worksite analysis, ensure physical protection, reinforce training, enforce discipline, and can explain work procedures.

(2) Supervisors received basic training in, appear to understand, and can show the importance of worksite analysis, physical protection, training reinforcement, discipline, and procedures.

(1) Supervisors make efforts to meet their safety/health responsibilities but have limited training.

(0) No formal effort to train supervisors in safety and health responsibilities is apparent.

Score: Comments:

IVB2: Leadership Education

(4) All managers receive formal hazard control education and understand all objectives.

(3) All managers follow and can explain their roles in hazard management control efforts.

(2) Managers generally show an understanding of their hazard control roles and responsibilities.

(1) Managers are generally able to describe their hazard control roles but often have trouble modeling it.

(0) Managers show little understanding of their safety and health management responsibilities.

Score: Comments:

Safety and Health Program Element	Possible Score	Actual Score
Management and Leadership	36	_____
Workplace Analysis	28	_____
Hazard Prevention and Control	24	_____
Education and Training	12	_____
Total	100	_____

Appendix B: Hazard Control Perception Survey

1. Did you receive adequate job-related training before assuming your current position?
2. Do supervisors discuss accidents, incidents, and injury events with involved workers?
3. Do supervisors enforce hazard control and safety rules fairly and correct unsafe behaviors?
4. Do supervisors take appropriate disciplinary action when work rules are not followed?
5. Do you perceive the major cause of accidents to be unsafe work conditions?
6. Does the organization actively promote and encourage employees to work safely?
7. Do you believe senior leaders view hazard control and safety as an important organizational priority?
8. Do supervisors seem more concerned with their personal records than with the accident?
9. Would some type of "safety incentive" program motivate you to work more safely?
10. Does the organization conduct required hazard surveys thoroughly?
11. Do you feel that supervisors receive adequate safety and health training?
12. Do supervisors provide sufficient training on the proper selection and use of personal protective equipment?
13. Do you understand what a performance-based occupational safety and health administration (OSHA) standard means?
14. Have you received any safety-related training or follow-up training since your orientation?
15. Does the organization keep records of safety inspections and identified hazards?
16. Do you feel that employees are influenced by the organizational hazard control efforts?
17. Does the organization provide you information about the costs, trends, types, and causes of accidents?
18. Do you feel the organization conducts accident investigations to assign blame?

19. Do you feel that the organization deals with problems caused by alcohol or substance abuse?
20. Does the organization conduct postaccident drug testing for all involved workers?
21. Do safety leadership personnel and supervisors conduct unscheduled hazard surveys and inspections?
22. Do you understand how the worker's compensation system works?
23. Does the organization provide special education and training for all shift workers?
24. Does the organization make safety a part of all job reviews or evaluations?
25. Do you think injured workers should participate in an "early return to work" initiative?
26. Is off-the-job safety an integral part of the overall hazard control and safety efforts?
27. Do supervisors report accidents promptly?
28. Do you wish to view the facility accident and injury record to compare it with similar facilities?
29. Do you feel your coworkers support the organization hazard control management efforts?
30. Do supervisors take hazard control and job safety seriously?
31. Does the organization take a proactive or reactive role when promoting hazard control?
32. Does the organization encourage supervisors to recognize those that work safely?
33. Do workers participate in the development of safe work practices and hazard control policies?
34. Do you feel workers play an important role in making hazard control and safety decisions?
35. Does senior management support the supervisor when they make decisions affecting hazard control?
36. Do coworkers understand the relationship between their job tasks and safety?
37. Do you know where to access the written emergency action plan?
38. Do you know where to access the hazard communication program?
39. Do you feel that you received appropriate hazard control orientation and training before beginning this job?
40. Do you feel that the company has too many rules governing safety and health issues?
41. Do supervisor-enforced hazard control and safety rules in same manner as other job-related policies?
42. Does the organization set hazard control or safety-related goals and objectives?
43. Does senior management communicate goals to all workers in the organization?

44. What role do employees take in the goal-setting process?
45. Who serves as the key person in the hazard control management function?
46. Do you feel the organization quickly evaluates hazards and takes appropriate actions?
47. Can supervisors reward workers for good safety performance?
48. Do you think alcohol and drugs increase the risk of an accident?
49. Do workers caution others about unsafe conditions and behaviors?
50. Can you initiate actions to correct an unsafe situation?
51. Do you know how to report an unsafe conditions, hazards, or behaviors?
52. Do workers fear the threat of reprisal when reporting safety deficiencies?
53. Do you feel hazard control and safety issues receive the same priority as other organizational issues?
54. Does the organization value its good compliance record over other hazard control objectives?
55. Do supervisors model safe behaviors to their workers?
56. Do supervisors promote safety with statements such as, "This is management's idea?"
57. Do all employees receive an adequate safety orientation?
58. Do you feel the safety orientation program adequately prepares you to work safely?
59. Does management recognize safety-related work behaviors?
60. Do safety meetings directly impact safety performance on the job?
61. Do workers have the opportunity to attend safety meetings and training classes?
62. Do supervisors handle workers with personal problems in an effective manner?
63. Do you view your job as stressful?
64. Do you know your organizational or departmental hazard control and safety goals?
65. Do supervisors consistently require use of personal protective equipment?
66. Can workers use alcohol or drugs on the job without detection?
67. Do supervisors sometimes overlook risks and hazards to get the job done?
68. Do hourly workers serve on the hazard control or safety committee?
69. Do you know the name of the organizational hazard control manager?
70. Does adherence with some established safety rules hinder job accomplishment?
71. Do you feel overworked?
72. Do you feel pushed on the job?
73. Does the organization mandate overtime work?

74. Do you feel satisfied with your job?
75. Do you feel that you can achieve the goals of your job?
76. Does the organization recognize you for doing a good job?
77. Do superiors assign you responsibility without delegating authority?
78. Do you feel you are overwhelmed with too much work-related responsibility?
79. Do you enjoy your job?
80. Does your immediate supervisor ask you for input?
81. Do you consider yourself loyal to the organization?
82. Do you feel any organization loyalty directed at you?
83. Do you believe in the importance of teamwork in your work area?
84. Does the organization provide you with adequate job-related training?
85. Do superiors ever judge you by things beyond your control?
86. Do you consider job security an important issue?
87. Do organizational hazard control plans, policies, and procedures protect the employee?
88. Do you believe that accidents will just happen?
89. Can the organizations truly prevent accidents?
90. Do workers feel free to discuss accident causal factors with investigators?
91. Do you feel that your job exposes you to more hazards than most other workers?
92. Do you understand the purpose of job-safety analysis?
93. Do you understand your responsibilities during an emergency or disaster?
94. Do you possess sufficient education and training to accomplish your job in a safe manner?
95. Do supervisors use safety practices and set an example for subordinates?
96. Do senior managers visit job areas and discuss the importance of working safely?
97. How would you classify hazard control efforts in your facility? Reactive___Proactive___
98. Do leaders and supervisor promote hazard control as the right thing to do, or do they promote compliance?
99. Does the organization require supervisors to conduct job-related safety training?
100. Do you know the most hazardous substance used in the workplace?
101. Do you know the two safest egress routes from your work area to a safe place?
102. Does the organization conduct realistic emergency egress drills on a regular basis?

103. What hazard or safety issue causes you the most concern?

_____

_____

_____

104. How would you improve the effectiveness of the hazard control management?

_____

_____

_____

105. What actions would you suggest for improving senior management involvement and employee participation in the organizational hazard control efforts?

_____

_____

_____

Appendix C: Sample Hazard Correction Form

Department/Committee: Date:
_____

Department or Hazard Location:
_____
_____

Method of Identification:
_____
_____

Hazard Description:
_____
_____
_____
_____
_____
_____
_____

Review Date	Plan of Action	Next Review Date

Description of Hazard Control Implemented_____

_____

_____

_____

_____

_____

Name of Person(s) Responsible for Action_____

_____

Follow-Up Evaluation Comments

_____

_____

_____

Final Resolution Date and Comments:

_____

_____

Appendix D: Sample Hazard Control Status Report

Facility/Department_____ Reporting Period_____

1. Brief Summary of Key Hazard Control and Safety Issues
2. Facility Security-Related Information
3. Orientation, Training, and Education Summary
4. Status of Hazard Control Policy and Procedural Reviews
5. Emergency Drills and Exercises
   a. Type and Scope
   b. Improvement Areas
6. Hazardous Material and Waste Issues
   a. Key Exposure Incidents
   b. New Hazardous Substances
7. Fire and Life Safety Issues
8. Information Collection and Evaluation System
   a. Key Accident Investigation Information
   b. Reports of Illnesses and Injuries
   c. Hazards Identified and Reported
9. Surveillance Activities
   a. Self-Inspections
   b. Hazard Surveys
   c. Compliance Inspections
   d. Insurance-Related Audits
10. Hazard Analysis and Control
    a. Controls and Corrections Implemented
    b. Major Trends Uncovered
    c. Job Hazard Analysis
11. Hazard Control–Related Equipment and Facility Issues
    a. Lockout/Tagout Procedures
    b. Periodic/Preventive Maintenance Issues
12. General Comments and Other Information

Appendix E: Hazard Control Improvement Principles

Organizations can apply the principles of quality improvement to hazard control efforts in a number of ways.

Develop a Policy: The organization should publish a hazard control policy that outlines objectives and goals. Leaders must communicate the policy to all members of the organization. Require everyone to participate and support hazard control efforts.

Promote the Importance of Inspections: Members at every organizational level must understand the purpose of hazard control inspections, audits, and survey.

Constantly Improve the Hazard Control Functions: Leaders, managers, and hazard control personnel must endorse the "philosophy" of continuous improvement.

Leadership and Management: Managers and supervisors must use effective management principles and leadership concepts to improve organizational efficiency. Top management must provide staff members with the tools and the time to pursue improvement ideas.

Promote Trust and Innovation: Provide a climate where organizational members at all levels can identify problems and make suggestions to improve operations. Trust creates an atmosphere that promotes innovation. Encourage coordination and open communication among all operational departments.

Eliminate Meaningless Slogans: Hazard control and safety messages must clearly define safe work practices, rules, or job procedures.

Promote Pride in Quality Work: Hazard control objectives must stress the importance of accomplishing jobs or tasks safely and correctly. Most individuals take pride in doing quality work. Provide organizational members with good equipment, a safe work area, and effective training.

Encourage Members to Work on Self-Improvement: Enhance organizational performance improvement efforts by affording everyone the opportunity to engage in educational and other personal-development opportunities.

Facilitate Change: Managers and supervisors must change or modify their management style to benefit the improvement of the organization. Leaders must not only accept change but also promote the need for change when warranted.

Appendix F: Accident Causal Factors Chart

Causal Factors	Identification of Factors	Possible Corrective Actions
Environmental: unsafe procedure or process	Hazardous processes; management failed to adequately plan	Job-hazard analysis or formulation of safe job practices
Defective, overused	Buildings, machines, or equipment worn, cracked, broken, or defective	Inspection; replacement; proper maintenance
Improperly guarded	Work areas, machines, or equipment that are unguarded or inadequately guarded	Inspection; check plans, blueprints, purchase orders, contracts, and materials; provide guards for existing hazards
Defective design	Failure to consider safety in design, construction, or installation of buildings, machinery, and equipment: too large, too small, or not strong enough	Unreliable source of supply; check plans, blueprints, purchase orders, contracts, and materials; provide guards for existing hazards
Unsafe dress or apparel	Management failed to provide/specify the use of goggles, respirators, safety shoes, hard hats, or safe apparel	Provide proper apparel or personal protective equipment; specify acceptable dress, apparel, or protective equipment
Unsafe housekeeping	Poor job area layout; lack of required equipment for good housekeeping— shelves, boxes, bins, aisle markers, etc.	Provide suitable layout and equipment necessary for good housekeeping

(Continued)

Causal Factors	Identification of Factors	Possible Corrective Actions
Improper ventilation	Poor ventilated or unventilated work areas	Improve the ventilation
Improper illumination	Poorly illuminated or no illumination at all	Improve the illumination
Behavior-related: lack of knowledge or skill	Unaware of safe practice; unskilled; not properly instructed or trained	Job training
Improper attitude	Worker properly trained but failed to follow instructions because of one of the following: willful, reckless, absent-minded, emotional, or angry	Supervisor; discipline; personnel work
Physical deficiencies	Poor eyesight, defective hearing, heart trouble, hernia, etc.	Preplacement physical examination; periodic physical examination; proper placement of employees; identification of workers with temporary bodily defects

Appendix G: Sample Ergonomic Symptom Report

NAME _____ DEPARTMENT _____

JOB TITLE _____ SHIFT _____

Identify All Affected Areas
   O Forearm   O Wrist   O Knee   O Elbow   O Upper Back
   O Lower Back   O Hand   O Fingers   O Ankle   O Foot
   O Shoulder   O Thigh   O Lower Leg   O Neck

Check Terms to Describe Complaint
   O Aching   O Numbness   O Tingling   O Loss of Color
   O Burning   O Swelling   O Weakness   O Stiffness
   O Cramping   O Other _____

1. List how and when the problem first occurred. _____
   _____

2. List the length and description of each episode. _____

3. How often did the problem occur during the past year? _____
   _____

4. What do you think caused the problem? _____
   _____

5. Describe the problem at its worst. _____
   _____

6 Describe any medical treatment for this problem. _____
   _____

7. What medical treatment helps the problem? _____
   _____

8. Describe frequency and type of treatment during the past year. ___
   _____

9. How many workdays did you lose because of this problem during the past year? _____

10. Have you been placed in another job or a modified duty status because of the problem?
_____
_____

11. What would help the problem? _____
_____

12. Do you work another job in addition to this one? Yes/No

13. If you answered "Yes" to Item 12, please describe the requirements of the job. _____
_____

14. List off-the-job activities, hobbies, or interests. _____
_____

15. What other did you work at during the past 12 months for more than two weeks?
_____

16. Provide any other helpful information. _____
_____

_____   _____

Name/Signature   Date

Appendix H: Hazardous Material Exposure Limits and Terms

Exposure Limit	Agency	Definition
TLV	ACGIH	Threshold limit value: The airborne concentration of a substance to which most workers can be exposed on a daily basis without adverse health effects.
TWA	ACGIH	Time-weighted average: The average concentration of a substance for a normal 8-hour workday and 40-hour workweek.
STEL	ACGIH OSHA	Short-term exposure limit: A 15-minute TWA that should not be exceeded at any time during the workday. There should be 60 minutes between each 15-minute exposure, up to four times a day.
CEILING (TLV-C)	ACGIH OSHA	Ceiling limit: The airborne concentration that is not to be exceeded at any time during the workday.
PEL	OSHA	Permissible exposure limit: Limit based on 8-hour TWAs. Exposures below the PEL do not require respiratory protection.
REL	NIOSH	Recommended exposure limit: A TWA for up to a 10-hour workday during a 40-hour work week.
IDLH	NIOSH	Immediately dangerous to life or health: Levels may cause severe health effects that may impair a person or prevent the ability to escape from a dangerous situation.

(Continued)

Exposure Limit	Agency	Definition
Acute exposure		Single exposure or several short-term exposures to a toxic substance.
Chronic exposure		Long-term exposure at a rate at which the body cannot get rid of the toxic substance.
Toxicity		A hazardous substance that poses a poison hazard to human health: • Immediate toxicity occurs rapidly after a single exposure episode. • Delayed toxicity occurs after a lapse of time. • Systematic toxicity is characterized by effects at a place other than the point of entry. • Local toxicity occurs when effects arise at the point of entry.

TLVs are given in parts per million (ppm) or milligrams per cubic meter of air (mg/m^3).
PEL values can be found in 29 CFR 1910.1000.

Appendix I: Hazardous Material Ratings

Numerical Hazard Ratings	
Health Hazards	
4	Deadly—the slightest exposure to this substance could be life-threatening. Only specialized protective clothing should be worn when working with this substance.
3	Extremely dangerous—serious injury could result from exposure to this substance. Do not expose any body surface to this material. Full protective measures should be taken.
2	Dangerous—exposure to this substance would be hazardous to health. Protective measures are indicated.
1	Slight health hazard—irritation or minor injury would occur from exposure to this substance. Protective measures are indicated.
Flammability	
4	Flash point below 73°F—substance is very volatile, explosive, or flammable. Use extreme caution when handling or storing the substance.
3	Flash point below 100°F—flammable, explosive, or volatile under most normal temperature conditions. Exercise great caution in using and storing a substance.
2	Flash point below 200°F—moderately heated conditions may ignite this substance. Use caution in handling or storage.
1	Flash point above 200°F—this substance must be preheated to ignite.

(Continued)

Numerical Hazard Ratings	
Reactivity	
4	May detonate—capable of explosion at normal temperatures. Evacuate area if exposed to heat or fire.
3	Explosive—substance is capable of explosion by strong initiating source such as heat, shock, or water.
2	Unstable—subject to violent chemical changes at normal or elevated temperatures. Potential violent explosive reaction may occur if exposed to water.
1	Normally stable—may become unstable at elevated temperatures.
Color Ratings	
Blue (Health hazard)	May cause health problems if acute exposure occurs by ingestion, inhalation, or physical contact.
Red (Flammability)	Evaluates the risk of materials to fire burst based on factors relative to the substance and surrounding environment.
Yellow (Reactivity)	Advises that a substance may react violently under certain conditions or exposures.
White (Specific hazard)	Refers to substances with specific hazards or properties, such as oxidizers.

Appendix J: Small-Facility HAZCOM Training Record

Employee Name:

Employer:

I, the undersigned, attended an organization-provided Hazard Communication Training Session that covered Phase I and II topics as indicated below.

Phase I Topics

1. Requirements and overview of the Hazard Communication Standard
2. Operations, locations, and processes with hazardous materials
3. The location and availability of the written HAZCOM plan
4. Methods and observations used to evaluate hazards, and how to detect the presence of hazardous materials in work areas
5. The session also covered the following topics:
    a. Description of labeling system
    b. Safety data sheet information
    c. Information on unlabeled pipes
    d. Hazards of nonroutine tasks
    e. How to obtain and use local hazard information

_____   _____

Employee Signature   Trainer's Initials/Date

Phase II Topics

1. The physical and health hazards of chemicals used in the department or work area
2. Measures workers must take to protect themselves from exposure

3. Specific procedures implemented to protect workers, such as the following:
   a. Safety rules
   b. Engineering controls
   c. Emergency response procedures
   d. Correct use of personal protective equipment

_____   _____

Employee Signature   Trainer's Initials/Date

Appendix K: Small-Facility Model Hazard Communication Plan

General

The following hazard communication (HAZCOM) plan has been established for this work site. This plan will be available for review by all employees during their work shift.

Hazard Determination

This work site relies on Safety Data Sheets (SDSs) obtained from product suppliers to meet the Occupational Safety and Health Administration (OSHA) hazard-determination requirements.

Labeling

A. The site supervisor will be responsible for ensuring that all containers entering the workplace are properly labeled according to the revised OSHA HAZCOM standard.
   1. All labels shall be checked for to determine identity of the material.
   2. Ensure the label contains appropriate hazard warnings.
   3. Verify that labels contain the name and address of the responsible party, such as manufacturer, distributor, or importer.
B. Each employee or supervisor shall be responsible for ensuring that all portable containers used in his or her work area are labeled with the appropriate material identity and hazard warnings.

Safety Data Sheets

A. The site supervisor will be responsible for compiling and maintaining the master SDS file. The file will be kept in/at the following location: _____
B. Additional copies of SDSs for employee use are located in/at the following location: _____
C. SDSs will be available for review to all employees during each work shift, and copies will be made available upon request to the site supervisor.

D. If a required SDS is not received with the shipment, the site supervisor shall contact the supplier, in writing, to request the Material Safety Data Sheet (MSDS).
E. If a SDS is not received after two such requests, the site supervisor should contact the local OSHA office for assistance in obtaining the SDS.

Employee Information and Training

A. The site supervisor shall coordinate and maintain records of employee HAZCOM training, including records of completion.
B. Before their initial work assignment, each new employee will be trained on the OSHA requirements under the standard. The training session will provide the following information and education:
   1. The requirements of the OSHA HAZCOM Standard.
   2. All operations in the new employee's work area where hazardous chemicals are present.
   3. The location and availability of the written HAZCOM plan, the list of hazardous chemicals, and the location of the SDSs.
   4. Methods, symptoms, and observations that can be used to detect the presence or release of hazardous chemicals in the work area.
   5. The physical and health hazards of the hazardous chemicals and measures the employees should take to protect from these hazards.
   6. Details of the HAZCOM program, including an explanation of labeling system and SDSs and how employees can obtain and use hazard information.
   7. The employee shall be informed that the employer is prohibited from discharging, or discriminating against, an employee who exercises his or her rights to obtain information regarding hazardous chemicals used in the workplace.
C. Before any new physical or health hazard is introduced into the workplace, each employee who may be exposed to the substance will be given information in the same manner as during the HAZCOM training class. Employees transferring to another department within the same organization will be trained on hazardous materials present in the new assignment.

Hazardous Nonroutine Tasks

Occasionally, employees may be required to perform nonroutine tasks. Before starting work, employees will be given information about the hazards of the area or procedure. This information will include: (1) specific chemical hazards, (2) protection/safety measures the employee can take to lessen risks of performing the task, and (3) measures the company has taken to eliminate or control the hazard. It is the policy of this work site that no employee will begin performance of a nonroutine task without first receiving appropriate safety and health training. Hazardous nonroutine tasks we have at our facility include: (List hazardous nonroutine tasks).

Multiemployer Work sites—Informing Contractors

A. If our company exposes any employee of another employer to any hazardous chemicals that we produce, use, or store, the following information will be supplied to that employer: (1) the hazardous chemicals their employees may encounter, (2) measures their employees can take to control or eliminate exposure to the hazardous chemicals, (3) the container and pipe labeling system used on-site, and (4) where applicable MSDSs can be reviewed or obtained.

B. Periodically, our employees may potentially be exposed to hazardous chemicals brought on our site by another employer. When this occurs, we will obtain from that employer information pertaining to the types of chemicals brought on-site, and measures that should be taken to control or eliminate exposure to the chemicals.

C. It is the responsibility of the site supervisor to ensure that such information is provided and/ or obtained before any services being performed by the contractor or vendor. To ensure that this is done, the following mechanism will be followed: (List all methods used to ensure the required information is provided or obtained).

Pipes and Piping Systems

Information on the hazardous contents of pipes and piping systems will be identified as necessary by label, sign, placard, or written operating instructions. Natural gas, steam, and compressed air lines should be identified if they pose a hazard to employees. Follow ANSI A13.1–1981 for appropriate color schemes.

List of Hazardous Chemicals

A list of all hazardous chemicals used at this work site is attached to this document. Further information regarding any of these chemicals can be obtained by reviewing its respective MSDS.

Materials which can be purchased by the ordinary household consumer, and which are used in the same fashion and amount as by the ordinary household consumer, are not required to be included in this list.

The list may be developed by category of chemical, alphabetical, or in any sequence deemed appropriate for the work site. Ensure employees understand how to use the hazardous chemical listing.

Hazardous Materials and Substances List

Hazardous Chemical (Use the same name as depicted on the container label and SDS).

Appendix L: NFPA Codes and Standards Listing

NFPA 1 Uniform Fire Code
NFPA 10 Standard for Portable Fire Extinguishers
NFPA 12 Standard on Carbon Dioxide Extinguishing Systems
NFPA 13 Standard for the Installation of Sprinkler Systems
NFPA 25 Standard for the Inspection, Testing, and Maintenance of Water-Based Fire Protection Systems
NFPA 30 Flammable and Combustible Liquids Code
NFPA 45 Standard on Fire Protection for Laboratories Using Chemicals
NFPA 50 Standard for Bulk Oxygen Systems
NFPA 51B Standard for Fire Prevention During Welding, Cutting, and Other Hot Work
NFPA 55 Standard for the Storage, Use, and Handling of Compressed Gases
NFPA 70 National Electrical Code 7
NFPA 70B Recommended Practice for Electrical Equipment Maintenance
NFPA 70E Standard for Electrical Safety Requirements for Employee Workplaces
NFPA 72 National Fire Alarm Code
NFPA 75 Standard for the Protection of Information Technology Equipment
NFPA 77 Recommended Practice on Static Electricity
NFPA 80 Standard for Fire Doors and Fire Windows
NFPA 82 Standard on Incinerators and Waste and Linen Handling Systems and Equipment
NFPA 85 Boiler and Combustion Systems Hazards Code

(Continued)

NFPA 90A Standard for the Installation of Air Conditioning and Ventilating Systems
NFPA 90B Standard for the Installation of Warm Air Heating and Air Conditioning Systems
NFPA 91 Standard for Exhaust Systems for Air Conveying of Vapors, Gases, Mists, and Particulate Solids
NFPA 92A Recommended Practice for Smoke Control Systems
NFPA 96 Standard for Ventilation Control and Fire Protection of Commercial Cooking Operations
NFPA 99 Standard for Health Care Facilities
NFPA 101 Life Safety Code
NFPA 101A Guide on Alternative Approaches to Life Safety
NFPA 101B Code for Means of Egress for Buildings and Structures
NFPA 105 Standard for the Installation of Smoke Door Assemblies
NFPA 110 Standard for Emergency and Standby Power Systems
NFPA 111 Standard on Stored Electrical Energy Emergency and Standby Power Systems
NFPA 115 Standard on Laser Fire Protection
NFPA 170 Standard for Fire Safety Symbols
NFPA 220 Standard on Types of Building Construction
NFPA 221 Standard for Fire Walls and Fire Barrier Walls
NFPA 230 Standard for the Fire Protection of Storage
NFPA 232 Standard for the Protection of Records
NFPA 434 Code for the Storage of Pesticides
NFPA 450 Guide for Emergency Medical Services and Systems
NFPA 471 Recommended Practice for Responding to Hazardous Materials Incidents
NFPA 472 Standard for Professional Competence of Responders to Hazardous Materials Incidents
NFPA 704 Standard System for the Identification of the Hazards of Materials for Emergency Response
NFPA 1600 Emergency Management and Business Continuity
NFPA 801 Standard for Fire Protection for Facilities Handling Radioactive Materials

NFPA 900 Building Energy Code
NFPA 1600 Standard for Disaster/Emergency Management and Business Continuity Programs
NFPA 1620 Recommended Practice for Preincident Planning
NFPA 1994 Standard on Protective Ensembles for Chemical/Biological Terrorism Incidents
NFPA 1999 Standard on Protective Clothing for Emergency Medical Operations
NFPA 5000 Building Construction and Safety Code

Appendix M: Occupational Safety and Health Administration Personal Protective Equipment Hazard Assessment Form

Department/Facility _____ Date _____

A. Eye and Face
 1) Flying Particles _____
 2) Molten Metal _____
 3) Liquid Chemicals _____
 4) Acids _____
 5) Caustic Liquids _____
 6) Chemical Gases or Vapors _____
 7) Light Radiation _____
 8) Other _____

B. Head
 1) Falling or Flying Objects _____
 2) Work Being Performed Overhead _____
 3) Elevated Conveyors _____
 4) Striking against a Fixed Object _____
 5) Forklift Hazards _____
 6) Exposed Electrical Conductors _____
 7) Other _____

C. Other
 1) Lifting _____
 2) Repetitive Motion, Prolonged Standing, and so on.
 3) Blood-Borne Pathogens _____

D. Foot
1) Falling and Rolling Objects _____
2) Objects Piercing the Sole _____
3) Electrical Hazards _____
4) Wet or Slippery Surfaces _____
5) Chemical Exposure _____
6) Environmental _____
7) Other _____

E. Hand
1) Skin Absorption _____
2) Cuts or Lacerations _____
3) Abrasions _____
4) Punctures _____
5) Chemical Burns _____
6) Thermal Burns _____
7) Harmful Temperature Extremes _____
8) Other

F. Respiratory
1) Harmful Dusts _____
2) Fogs _____
3) Fumes _____
4) Mists _____
5) Smokes _____
6) Sprays _____
7) Vapors _____
8) Other _____

G. Torso
1) Hot Metals _____
2) Cuts _____
3) Acids _____
4) Radiation _____

H. Comments _____
_____
_____
_____
_____
_____

I. Certification

The assessment used the following methods to determine existence of workplace hazard requiring PPE:

- Walk-through survey
- Specific job analysis

- Review of accident statistics
- Review of safety equipment selection guideline materials
- Selection of appropriate or required PPE

Assessment Conducted by _____ Date _____

Assessment Certified by _____ Date _____

Appendix N: Sample Initial Accident Investigation Form

Facility/Department _____

Date of Event _____ Date Reported _____

Names of Persons Involved _____

Time of Accident _____ Location of Accident _____

Name of Supervisor _____

Machines/Tools/Processes/Operations Involved _____

_____

Brief Description of Injuries _____

_____

Property Damage _____

Witnesses _____

_____

Contributing Causes (Check Each Appropriate Causal Factor)

[] Improper instruction [] Improper maintenance [] Unsafe position [] Physical impairment [] Lack of training [] Human error [] Poor ventilation [] Improper clothing [] Horseplay [] Lack of supervision [] Improper procedure [] No authority to operate [] Unsafe arrangement [] Failure to use PPE [] Unsafe equipment [] Poor housekeeping [] Failure to secure [] Inoperative safety device [] Improper guarding [] Safety rule violation [] Using wrong tool [] Failure to lockout [] Other (Describe) _____

_____

Summary of Initial Investigation _____

_____

Corrective Actions Taken _____

_____

Other Comments _____

_____

Supervisor Signature and Date _____

Appendix O: Small-Facility Model Respirator Plan

General Information

The purpose of this respirator plan is to protect all employees from respiratory hazards through the effective use of respirators. The Respirator Plan Administrator (RPA) appointed at this location is: _____. The employer has expressly authorized the RPA to audit and change respirator usage procedures whenever there is a chance of exposure to an air contaminant or airborne disease at the work site. This authority includes designating mandatory respirator usage areas and/or job-related tasks. The RPA is solely responsible for all aspects of this plan and has full authority to make decisions relevant to respirator usage. This authority includes training workers, purchasing the necessary equipment to implement the program, and developing local respiratory protection procedures.

Local Written Operating Practices

The RPA will develop written "operating practices" that detail specific instructions covering the basic elements in this plan. These local operating practices will be attached to the plan. These practices can only be amended by the RPA.

The RPA will develop detailed written standard operating procedures governing the selection and use of respirators, using the Occupational Safety and Health Administration (OSHA) standard and the National Institute of Occupational Safety and Health (NIOSH) Respirator Decision Logic as guidelines. Outside consultation, manufacturers assistance, and other recognized authorities will be consulted if there is any doubt regarding proper selection and use of respirators. These detailed operating practices will be included as attachments to this respirator plan.

Respirators Selection

Respirators will be selected on the basis of exposures. All selections will be made by the RPA and only NIOSH certified respirators will be selected and used.

Training and Use

Respirator users will be instructed and trained in the proper use of respirators and their limitations. Both supervisors and workers will be trained by the RPA. The training should provide the employee an opportunity to handle the respirator, have it fitted properly, test its face piece-to-face seal, wear it in normal air during a familiarity period, and finally to wear it in a test atmosphere. Fit testing will be accomplished for all tight fitting respirators used at this location.

Fitting Instructions

Every respirator wearer will receive fitting instructions, including demonstrations and practice in how the respirator should be worn, how to adjust it, and how to determine if it fits properly. Respirators should not be worn when conditions prevent a good face seal. Such conditions may be a growth of beard, sideburns, a skull cap that projects under the face piece, or temple pieces on glasses. No employees of this facility, who are required to wear tight fitting respirators, may wear beards. Also the absence of one or both dentures can seriously affect the fit of a face piece. The workers diligence in observing these factors will be evaluated by periodic checks. To assure proper protection, the user seal check will be done by the wearer each time she/he puts on the respirator. The manufactures instructions will be followed.

Assignment of Respirators

When practicable, the respirators will be assigned to individual workers for their exclusive use. Nondisposable respirators will be regularly cleaned and disinfected. Those issued for the exclusive use of one worker will be cleaned after each days use, or more often if necessary. Those used by more than one worker will be thoroughly cleaned and disinfected after each use. The RPA will establish a respirator cleaning and maintenance facility and develop detailed written cleaning instructions. Disposable respirators will be discarded, if they are soiled or are no longer functional. For additional information, refer to the manufacturer's instructions. Respirators used routinely will be inspected during cleaning. Worn or deteriorated parts will be replaced.

Employee Screening

Persons will not be assigned to tasks requiring use of respirators unless it has been determined that they are physically able to perform the work and use the respirator. The employer will designate a health care professional to determine and assess worker health and physical ability to wear a respirator. The respirator user's medical status will be reviewed annually.

Evaluations

The employer will ensure that an annual inspection/evaluation of the program is conducted to determine the continued effectiveness of the program. The RPA will make frequent inspections of all areas where respirators are used to ensure compliance with the respiratory protection requirements.

Evaluation Checklist

In general, the respiratory protection plan should be evaluated for each job or at least annually, with program adjustments, as appropriate, made to reflect the evaluation results. Functions can be separated into administration and operation.

Administration

1. Is there a written policy that acknowledges employer responsibility for providing a safe and healthful workplace, and assigns program responsibility, accountability, and authority?
2. Is program responsibility vested in one individual who is knowledgeable and who can coordinate all aspects of the program at the health care facility?
3. Can administrative and engineering controls eliminate the need for respirators?
4. Are there written procedures/statements covering the following topics?
   _____ a. designation of an administrator
   _____ b. respirator selection
   _____ c. purchase of NIOSH certified respirators
   _____ d. medical aspects of respirator usage
   _____ e. issuance of equipment
   _____ f. fitting
   _____ g. training
   _____ h. maintenance, storage, and repair
   _____ i. inspection
   _____ j. use under special conditions
   _____ k. work area surveillance

Operation
5. **Equipment Selection**
   a. Are work area conditions and worker exposures properly surveyed?
   b. Are respirators selected on the basis of the hazard of exposure?
   c. Are selections made by individuals knowledgeable in selection procedures?
6. Are only NIOSH certified respirators purchased and used? Do they provide adequate protection for the specific hazard?
7. Has a medical evaluation of the prospective user been made to determine physical and psychological ability to wear the selected respiratory protective equipment?
8. Where practical, have respirators been issued to the users for their exclusive use, and are there records covering issuance?

9. **Respiratory Protective Equipment Fitting**
   a. Are the users given the opportunity to try on several respirators to determine whether the respirator they will be subsequently wearing is the best fitting one?
   b. Is the fit tested at appropriate intervals?
   c. Are those users who require corrective lenses properly fitted?
   d. Is the face piece tested in a test atmosphere?
   e. Are workers prohibited from wearing respirators in contaminated work areas when they have facial hair or other characteristics which may cause face seal leakage?

10. **Respirator Use in the Work Area**
    a. Are respirators being worn correctly (i.e., head covering over respirator straps)?
    b. Are workers keeping respirators on all the time while in the designated areas?

11. **Maintenance of Respiratory Protective Equipment**
    a. Are nondisposable respirators cleaned and disinfected after each use when different people use the same device, or as frequently as necessary for devices issued to individual users?
    b. Are proper methods of cleaning and disinfecting utilized?
    c. Are respirators stored in a manner so as to protect them from dust, sunlight, heat, damaging chemicals, or excessive cold or moisture?
    d. Are respirators stored in a storage facility so as to prevent them from deforming?
    e. Is storage in lockers permitted only if the respirator is in a carrying case or carton?

12. **Inspection**
    a. Are respirators inspected before and after each use and during cleaning?
    b. Are qualified individuals/users instructed in inspection techniques?
    c. Are records kept of the inspection of respiratory protective equipment?

13. **Repair**
    a. Are replacement parts used in repair those of the manufacturer of the respirator?
    b. Are repairs made by trained individuals?

14. **Training and Feedback**
    a. Are users trained in proper respirator use, cleaning, and inspection?
    b. Are users trained in the basis for selection of respirators?
    c. Are users evaluated, using competency-based evaluation, before and after training?
    d. Are users periodically consulted about program issues such as discomfort, fatigue, etc.?

Sample Respirator Inspection Record

TYPE _____

NO. _____ DATE: _____

A. Face piece
B. Inhalation Valve
C. Exhalation Valve Assembly
D. Headbands/Straps
E. Filter Cartridge
F. Cartridge/Canister
G. Harness Assembly
H. Hose Assembly
I. Speaking Diaphragm
J. Gaskets
K. Connections
L. Other Defects

DEFECTS FOUND:

CORRECTIVE ACTION:

Medically Screen All Users

Conduct a medical evaluation of workers to determine fitness to wear to wear respirators. The use of respirators can place several physiological stresses on wearers–stresses that particularly involve the pulmonary and cardiac systems. However, respirators typically used by health care workers are generally lightweight, and the physiological stresses they create are usually small. Therefore, most workers can safely wear respirators. Current OSHA regulations (29 CFR 1910.139) state that workers should not be assigned tasks requiring respirators unless they have been determined to be physically able to perform the work while using the equipment. The regulations also note that a physician should determine the criteria on which to base this determination.

No general consensus exists about what elements to include in medical evaluations for respirator use in general industry. Some institutions use only a questionnaire as a screening tool; others routinely include a physical examination and spirometry; and some include a chest X-ray. No generally accepted criteria exist for excluding workers from wearing respirators. Specifically, no spirometric criteria exist for exclusion. However, several studies have shown that most workers with mild pulmonary function impairment can safely wear respirators. There are some restrictions, such as the type of respirator or workload, for those with moderate impairment. There should be no respirator wear for individuals with severe impairment. Some respirators have a latex component and should not be worn by those who are allergic to latex.

Because most health care workers wear the very light, disposable half-mask respirator, recommend that a health questionnaire be the initial step in the evaluation process. Refer to OSHA 29 CFR 1910.134 paragraph "e" for guidance on medical evaluation. Appendix B of the Standard contains a sample medical questionnaire. If results from this evaluation are essentially normal, the employee can be cleared for respirator wear. Further evaluation, possibly including a directed physical examination and/or spirometry, should be considered in cases in which potential problems are suggested on the basis of the questionnaire results.

Appendix P: Sample Workplace Violence Prevention Policy

Policy Statement

The safety and security of organization personnel, patients, and visitors is of vital importance. Threats, threatening behavior, or acts of violence against personnel, patients, visitors, or contractors will not be tolerated. It is the policy of the organization to provide a safe environment to conduct the mission of the organization in the most effective manner possible. This policy supports the written procedures set forth by the organization's safety and health workplace violence plan, which is incorporated into this policy by reference. A safe environment will be attained by the following:

1. Appropriate employee screening
2. Employee education and training
3. Surveillance of the work area
4. Effective management of situations involving violence or threats of violence

Definitions

Workplace: Any location, either permanent or temporary, wherein an employee performs any work-related duty. This includes, but is not limited to, the building or facility and the surrounding perimeters, including the parking lots, field locations, alternate work locations, and travel to and from work assignments.

Workplace violence: Any physical assault, threatening behavior, or verbal abuse by employees or third parties that occurs in the workplace. It includes, but is not limited to, beating; stabbing; suicide; attempted suicide; shooting; rape; psychological trauma such as threats, obscene phone calls, and intimidating presence; and harassment of any nature, such as stalking, shouting, or swearing.

Procedures

The organization will not tolerate the following conduct or behavior:

1. Threats, direct or implied
2. Physical conduct that results in harm to people or property
3. Possession of weapons on any property or workplace of the organization
4. Intimidating conduct or harassment that results in fear for personal safety

Inappropriate and threatening behaviors include, but are not limited to, the following:

1. Unwelcome name-calling, obscene language, and other verbally abusive behavior
2. Throwing objects, regardless of the type or whether a person is a target of a thrown object
3. Touching a person in an intimidating, malicious, or sexually harassing manner
4. Acts such as hitting, slapping, poking, kicking, pinching, grabbing, and pushing
5. Physical and intimidating acts such as obscene gestures, getting in your face, and fist shaking

Reporting and Investigating

Any employee who experiences, observes, or has knowledge of actual or threatened workplace intimidation or violence has a responsibility to report the situation as soon as possible:

1. In the case of an actual or imminent act or threat of violent behavior.
2. In all cases, the report should be immediately made to the employee's supervisor or department head and to the compliance committee.

All reports of workplace intimidation or violence will be investigated impartially and as confidentially as possible.

Employees are required to cooperate in any investigations. A timely resolution of each report should be reached and communicated to all parties involved as soon as possible.

Any form of retaliation against employees for making a bona fide report concerning workplace intimidation or violence is prohibited and must be immediately reported to the compliance committee.

Reporting Non-Work-Related Violence

Employees who are victims of domestic or non-work-related violence, or who believe they are potential victims of such violence, or who believe they are potential victims of such violence and fear it may enter the workplace, are encouraged to promptly notify their supervisor or department head. All such reports will be investigated as described earlier.

Non-disciplinary and Disciplinary Action

Upon completion of an investigation, incidents will be reviewed before proceeding with non-disciplinary or disciplinary action, according to the provisions of the performance and corrective-action policy.

Examples of actions that might be taken when an employee has been found to have violated the policy include, but are not limited to, the following:

1. Mandatory participation and counseling
2. Corrective action up to and including termination
3. Criminal arrest and prosecution
4. Initiation of a court order

Those who believe they are victims of intimidation or violence, whether workplace or non-work-related, should contact their supervisor or department head to inquire about available employee assistance and to obtain advice about dealing with the situation.

Weapons Policy

The possession, carrying, or use of weapons on the organization's property is strictly prohibited. This includes firearms, edged weapons, illegal knives, martial arts weapons, clubs, and any device capable of projecting a ball, pellet, arrow, bullet, shell, or other similar devices. Violation of this policy is grounds for immediate termination. Refer to the firearms and other weapons policy, which is incorporated into this policy by reference.

Appendix Q: Hazard Control–Related Acronyms

AAS: atomic absorption spectroscopy
ABC: airway, breathing, circulation
ABIH: American Board of Industrial Hygiene
ABSA: American Biological Safety Association
ACBM: asbestos-containing building material
ACCSH: Advisory Committee on Construction Safety and Health
ACGIH: American Conference of Governmental Industrial Hygienists
ACM: asbestos-containing material
ACP: area contingency plan
ACRSP: Association of Canadian Registered Safety Professionals
ACS: American Chemical Society
ADA: Americans with Disabilities Act
ADR: Alternative Dispute Resolution
AE: atomic emission
AEL: acceptable exposure limit
AGA: American Gas Association
AGST: aboveground storage tank
AHM: acutely hazardous material
AIChE: American Institute of Chemical Engineers

(Continued)

AIDS: acquired immune deficiency syndrome
AIHA: American Industrial Hygiene Association
ALARA: as low as reasonably achievable
ALJ: administrative-law judge
ALS: advanced life support
ANPR: advanced notice of proposed rule-making
ANSI: American National Standards Institute
APF: assigned protection factor
APHA: American Public Health Association
API: American Petroleum Institute
APIH: Association of Professional Industrial Hygienists.
APR: air-purifying respirator
ASHRAE: American Society of Heating, Refrigerating, and Air Conditioning Engineers
ASME: American Society of Mechanical Engineers
ASP: associate safety professional
ASQC: American Society for Quality Control
ASSE: American Society of Safety Engineers
AST: aboveground storage tank
ASTM: American Society for Testing Materials
ATC: automatic temperature compensation
ATCM: air-toxics control measure
ATSDR: Agency for Toxic Substances and Disease Registry
AWT: advanced wastewater treatment
BACT: best available control technology
BBP: blood-borne pathogens
BBS: behavior-based safety
BCHCM: Board of Certified Hazard Control Management
BCSP: Board of Certified Safety Professionals
BCT: best conventional pollutant-control technology
BLS: basic life support
BLS: Bureau of Labor Statistics

BPT: best practicable control technology currently available (CCAA)
BS: British standards
BSC: biological safety cabinet
BSI: British standards institute
BTU: British thermal unit
CAA: Clean Air Act
CAIR: Comprehensive Assessment Information Rule
CAMEO: computer-aided management of emergency operations
CAP: College of American Pathologists
CAS: Chemical Abstracts Service
CAV: constant air volume
CCP: Cooperative Compliance Program
CCPS: Center for Chemical Process Safety
CDC: Centers for Disease Control and Prevention
CEPP: chemical emergency-preparedness program
CEPS: cluster environmental protection specialist
CERCLA: Comprehensive Environmental Response, Compensation, and Liability Act
CET: certified environmental trainer
CFC: chlorofluorocarbon
CFM: cubic feet per minute
CFR: Code of Federal Regulations
CGA: Compressed Gas Association
CGL: comprehensive general liability insurance
CHCM: certified hazard control manager
CHEMTREC: Chemical Transportation Emergency Center
CHEP: certified health-care emergency professional
CHCM-SEC: certified hazard control manager, security
CHMM: certified hazardous materials manager
CHO: chemical hygiene officer
CHP: certified health professional
(Continued)

CHP: certified health physicist; chemical hygiene plan
CHSP: certified health-care safety professional
CHSP-FSM: certified health-care safety professional–fire safety manager
CHST: construction safety and health technician
CIH: certified industrial hygienist
CIIT: Chemical Industry Institute of Technology
CPC: chemical protective clothing
CPL: compliance directive
CPR: cardiopulmonary resuscitation
CPSC: Consumer Products Safety Commission
CPSM: certified product safety manager
CPSO: certified patient safety officer
CR: case report number
CSGs: clinical service groups
CSP: certified safety professional
CTD: cumulative trauma disorder
CWA: Clean Water Act
dBA: decibels (A scale)
DES: Department of Emergency Services
DHS: Department of Homeland Security
DHHS: Department of Health and Human Services
DM: dust and mist
DNR: Department of Natural Resources
DO: dissolved oxygen
DOD: Department of Defense
DOE: Department of Energy
DOI: Department of the Interior
DOJ: Department of Justice
DOL: Department of Labor
DOT: Department of Transportation
DVO: diffuse viewing only

EAP: emergency action plan
EAP: employee assistance program
ED: emergency department
EDP: electronic data processing
EEOC: Equal Employment Opportunity Commission
EHS: environment, health, and safety; or environmental health and safety
EHS: extremely hazardous substance
EIS: Environmental Impact Statement (NEPA)
ELF–EMF: extremely low frequency electric and magnetic fields
EM: emergency management
EMA: Emergency Management Agency
EMR: electromagnetic radiation
EMS: emergency management services; emergency medical service
EOC: emergency operations center
EOCS: environment of care standards
EP: extraction procedure
EPs: elements of performance
EPA: Environmental Protection Agency
EPCRA: Emergency Planning and Community Right-to-Know Act
ERA: environmental risk assessment
ERC: emissions reduction credit
ERG: emergency response guide
ERP: emergency response plan
ERPG: emergency response planning guideline
ERRIS: Emergency and Remedial Response Inventory System
ERT: emergency response team
ES&H: environment, safety, and health
ESA: Endangered Species Act
ESA: environmental site assessment
ESC: evidence of standards compliance

(Continued)

ESCBA: escape self-contained breathing apparatus
EtO: ethylene oxide
FAQs: frequently asked questions
FBI: Federal Bureau of Investigation
FDA: Food and Drug Administration
FEMA: Federal Emergency Management Agency
FIFRA: Federal Insecticide, Fungicide, and Rodenticide Act
FM: Factory Mutual; fire marshal
FMCSR: Federal Motor Carrier Safety Regulations
FMEA: failure mode and effect analysis
FMECA: failure mode, effects, and criticality analysis
FOIA: Freedom of Information Act
FR: flame resistant
FR: *Federal Register*
FS: feasibility study
FSAR: Final Safety Analysis Report
FTA: fault-tree analysis
FTU: fixed treatment unit
FWPCA: Federal Water Pollution Control Act
GACT: generally available control technology
GAO: General Accounting Office
GC/MS: gas chromatography/mass spectrometry
GERT: general employee radiation training
GFCI: ground-fault circuit interrupter
GLC: ground-level concentration
GLP: good laboratory practice
gpm: gallons per minute
GSA: General Service Administration
GW: groundwater
HAPs: hazardous air pollutants
HASP: health and safety plan

HazCom: hazard communication
HAZMAT: hazardous materials
HAZWOPER: hazardous waste operations and emergency response
HBV: hepatitis B virus
HCA: Hazard Communication Act
HCFCs: hydrochlorofluorocarbons
HCl: hydrogenchloride
HCO: health-care organization
HCS: hazard communication standard
HEPA: high-efficiency particulate air (filtration)
HF: hydrogen fluoride
HID: high-intensity discharge
HIV: human immunodeficiency virus
HMEP: hazardous-materials emergency preparedness
HMIG: hazardous-materials identification guide
HMIS: hazardous materials information system
HMRT: hazardous-materials response team
HMTC: hazardous-materials technical center
HP: health physicist
HRS: hazard ranking system
HS: hazardous substance
HSWA: Hazardous and Solid Waste Amendments
HVAC: heating, ventilation, and air conditioning
HWMU: hazardous-waste management units
IAFF: International Association of Firefighters
IAP: incident action plan
IAQ: indoor air quality
IARC: International Agency for Research on Cancer
IATA: International Air Transport Association
IBNR: incurred but not reported

(Continued)

IBR: incorporated by reference
IC: incident command
ICAO: International Civil Aviation Organization
ICC: Interstate Commerce Commission
ICP: incident command post
ICS: incident command system
IDLH: immediately dangerous to life and health
IEC: International Electrotechnical Commission
IHMM: Institute of Hazardous Materials Management
IR: infrared
ISEA: Industrial Safety Equipment Association
ISO: International Organization for Standardization
IUPAC: International Union of Pure and Applied Chemistry
JC: joint commission
JHA: job hazard analysis
JIC: joint information center
LCD: liquid crystal display
LEA: local enforcement agency
LEC: local emergency coordinator
LED: light-emitting diode
LEL: lower explosive limit
LEPC: local emergency planning committee
LIMS: laboratory information management system
LSO: laser safety officer
LUST: leaking underground storage tank
LWDII: lost workday due to injury and illness
MAWP: maximum allowable working pressure
MCL: maximum concentration limits or maximum contaminant level
MCS: multiple chemical sensitivity
MECO: Modern Engineering Company
mg/m3: milligrams per cubic meter

mil: 1 mil = 1/1000 of an inch
MIS: management information systems
mL: milliliter (also ml)
MMAD: mass median aerodynamic diameter
MOA: memorandum of agreement
MOS: measures of success
MOU: memorandum of understanding
mppcf: million particles per cubic foot
MS: mass spectroscopy
MSDS: Material Safety Data Sheet
MTD: maximum tolerated dose
MTU: mobile treatment unit
MUTCD: *Manual for Uniform Traffic Control Devices*
NAICS: North American Industry Classification System
NAS: National Academy of Sciences
NCP: National Contingency Plan
NCP: National Oil and Hazardous Substances Pollution Contingency Plan
NCRIC: National Chemical Response and Information Center
NDPES: National Pollutant Discharge Elimination System
NEIC: National Enforcement Investigations Center
NEMA: National Electrical Manufacturers Association
NEPA: National Environmental Policy Act
NESHAPs: National Emission Standards for Hazardous Air Pollutants
NFC: National Fire Code
NFPA: National Fire Protection Association
NFR: National Fire Rating
NH_3: ammonia
NHTSA: National Highway Traffic Safety Administration
NiCad: nickel–cadmium
NIEHS: National Institute of Environmental Health Sciences

(Continued)

NIH: National Institutes of Health
NIHL: noise-induced hearing loss
NIIMS: National Interagency Incident Management System
NIOSH: National Institute of Occupational Safety and Health
NIST: National Institute of Standards and Technology
NMFC: National Motor Freight Class
NMR: nuclear magnetic resonance spectroscopy
NOAA: National Oceanic and Atmospheric Administration
NPDES: National Pollutant Discharge Elimination System
NPL: National Priority List
NRC: Nuclear Regulatory Commission; National Response Center
NRDA: Natural Resource Damage Assessment
NRDAR: Natural Resource Damage Assessment and Restoration
NRR: noise-reduction rating
NRS: National Response System
NRT: National Response Team
NRTL: nationally recognized testing laboratory
NSC: National Safety Council
NSF: National Sanitation Foundation
NSF: National Science Foundation
NSFCC: National Strike Force Coordination Center
NTIS: National Technical Information Service
NTP: National Toxicology Program
NWPA: Nuclear Waste Policy Act
OBES: Office of Basic Energy Sciences
OCMV: open container in motor vehicle
OECM: Office of Enforcement and Compliance Monitoring
OMB: Office of Management and Budget
OPA: Oil Pollution Act of 1990
ORM: other regulated material
ORR: operational readiness review

OSC: on-scene coordinator
OSHA: Occupational Safety and Health Administration
OSHRC: Occupational Safety and Health Review Commission
OSWER: Office of Solid Waste and Emergency Response
OTA: Office of Technology Assessment
OV/AG: organic vapor/acid gas
PAPR: powered air-purifying respirator
PCB: polychlorinated biphenyls
PE: professional engineer
PEL: permissible exposure limit
PFA: priority focus areas
PFP: priority focus process
PFS: Professional Food Systems
PHA: preliminary hazards analysis
PHA: process hazards analysis
PHS: particularly hazardous substance
PIH: poison inhalation hazard
PL: public law
POA: plan of action
POP: performance-oriented packaging
POTW: publicly owned treatment works
ppb: parts per billion
PPE: personal protective equipment
ppm: parts of contaminant per million parts of air or fluid
PPR: periodic performance review
PRCS: permit-required confined space
PRP: potentially responsible party
PSA: preliminary site assessment
PSEL: plant-specific emission limit
PSI: pollution standards index

(Continued)

psi: pounds-per-square-inch
psig: pounds-per-square-inch gauge
PSM: process safety management
PVA: polyvinyl alcohol
QA/QC: quality assurance/quality control
QRA: quantitative risk assessment
R&D: research and development
RCRA: Resource Conservation and Recovery Act
RFI: radio-frequency interference
RFI: requirement for improvement
RIH: registered industrial hygienist
RMI: repetitive-motion injury
RMP: risk management program
rms: root mean squared
RP: responsible party
RRT: regional response team
RSI: repetitive strain injury
RSO: radiological safety officer or radiation safety officer
RSPA: Research and Special Programs Administration
RTK: right to know
SARA: Superfund Amendments and Reauthorization Act
SATA: site assessment and technical assistance
SCBA: self-contained breathing apparatus
SDWA: Safe Drinking Water Act
SEI: Safety Equipment Institute
SERC: State Emergency Response Commission
SHEM: safety, health, and environmental management
SHEP: safety, health, and environmental program
SIC: standard industrial classification
SIP: State Implementation Plan
SITE: Superfund Innovative Technology Evaluation

SOP: standard operating procedure
SPCC: spill prevention: control: and countermeasures
SQG: small-quantity generator
STEL: short-term exposure limit
STP: standard temperature and pressure
SUD: safe-use determination
SWA: Solid Waste Act
SWMP: Storm Water Monitoring Program
SWMU: Solid Waste Management Unit
SWPPP: Storm Water Pollution Prevention Program
TCLP: toxicity characteristic leaching procedure
TDS: totally dissolved solid
TLSI: the Laboratory Safety Institute
TLV: threshold limit value
TPQ: threshold planning quantity
TQM: Total Quality Management.
TSCA: Toxic Substance Control Act
TSDF: treatment: storage: and disposal facility
TSI: Transportation Safety Institute
TSR: technical safety requirement
TWA: time-weighted average
UBC: Uniform Building Code
UC: unified command
UEL: upper explosive limit
UFC: Uniform Fire Code
UGST: underground storage tank
UGT: underground tank
UL: Underwriters Laboratories
UM: Uniform Manifest
USC: United States Code

(Continued)

USCG: United States Coast Guard
USDA: US Department of Agriculture
USDW: underground source of drinking water
USEPA: United States Environmental Protection Agency
USFDA: United States Food and Drug Administration
USM: United States Marshal
USP: United States Pharmacopoeia
UST: underground storage tank
UV: ultraviolet
VDT: video display terminal
VGA: video graphics array
VHAP: volatile hazardous air pollutant
VOC: volatile organic compound
VPP: voluntary protection program
WBGT: wet bulb globe temperature
WEEL: workplace environmental exposure limit
WHMIS: workplace hazardous materials information system
WHO: World Health Organization
WMD: weapons of mass destruction
WPS: worker protection standard
WWTP: wastewater treatment plant

Appendix R: Hazard Control Glossary of Terms

A

Absorption: Transformation of radiant energy into a different form of energy by the interaction of matter, depending on temperature and wavelength or the process by which a liquid penetrates the solid structure of absorbents, fibers or, particles.

Absorbed dose: The amount of energy deposited by ionizing radiation in a unit mass of tissue.

Accessible: An ADA term about having the legally required features and/or qualities that ensure entrance, participation, and usability of places, programs, services, and activities by individuals with a wide variety of disabilities.

Accessible emission limit: The maximum accessible emission level permitted within a particular laser class.

Accessible emission level: The magnitude of accessible laser or collateral radiation of a specific wavelength or emission duration at a particular point as measured by appropriate methods.

Accident type: The description and classification of a mishap.

Acid: A compound either inorganic or organic that (1) reacts with a metal to evolve hydrogen, (2) reacts with a base to form a salt, (3) dissociates in water solution to yield hydrogen ions, (4) has a pH of less than 7, and (5) neutralizes bases or alkaline media by receiving a pair of electrons from the base, so that a covalent bond is formed between the acid and the base.

Action level: The amount of a material in air at which certain OSHA requirements take effect for the protection of workers. Exposure at or above the action level is termed an "occupational exposure."

Action plan: Documented outline of specific projected activities to be accomplished within a specified period, to meet a defined need.

Active electrode: Electrosurgical accessory that directs current flow to the surgical site—also called a cautery tip.

Activity (radioactivity): The rate of decay of radioactive material expressed as the number of atoms breaking down per second, measured in units called becquerels or curies.

Actuator: A power mechanism used to effect motion of a robot or a device that converts electrical, hydraulic, or pneumatic energy into robot motion.

Acute effect: An adverse effect on humans or animals, with symptoms developing rapidly and quickly becoming a crisis, resulting from a short-term exposure.

Acute radiation exposure: An exposure to radiation that occurred in a matter of minutes rather than in longer or continuing exposure over a period of time.

Acute radiation syndrome: A serious illness caused by receiving a dose of more than 50 rads of penetrating radiation to the body in a short time (usually minutes); the early symptoms include nausea, fatigue, vomiting, and diarrhea; loss of hair, swelling of the mouth and throat, and general loss of energy may follow.

Adaptation: A change in the structure of an organism, which results in its adjustment to its surroundings.

Adequate: Denotes the quality or quantity of a system, process, procedure, or resource that will achieve the relevant incident response objective.

Adsorption: Attachment of the molecules of a gas or liquid to the surface of another substance, normally a solid, or the process by which a liquid adheres to the surface of a material but does not penetrate the fibers of the material.

Aerobic: Requiring the presence of air or oxygen to live, grow, and reproduce.

Aerosol: Liquid droplets or solid particles dispersed in air.

Agency: A division of government with a specific function offering a particular kind of assistance or in the emergency incident command system.

Air exchange rate: Speed at which outside air replaces air inside a building, or the number of times the ventilation system replaces air within a room or built structure.

Airborne radioactive material: Any radioactive material dispersed in the air in the form of a dust, fumes, mists, aerosols, vapors, or gases.

Alcohol-based hand sanitizers: A gel or rub that contains alcohol to reduce the number of viable microorganisms on the hands (60%–95% ethanol or isopropanol).

Alkali: A term normally used to refer to hydroxides and carbonates of the metals of group 1a of the periodic table, or ammonium hydroxide.

All hazards: Emergency management term referring to a natural or man-made event that would require actions to protect life, property, environment, public health, safety, and/or minimize disruptions to government, social, or economic activities.

Area command: An organization established to oversee the management of multiple incidents that are each being handled by separate incident command systems or organizations, to oversee the management of a very large or evolving incident that has multiple incident management teams engaged, an area command is activated only if necessary, depending on the complexity of the incident and incident management span-of-control considerations.

Allergen: A substance or particle that causes an allergic reaction.

Alloy: A mixture or solution of metals, either solid or liquid, which may or may not include a nonmetal.

Alpha particle: Positively charged particle emitted by certain radioactive materials; it is identical to the nucleus of the helium atom and is the least penetrating form of radiation.

Alternate site burn: A patient burn resulting from electricity exiting the body by unintended means.

Ambient air: Outside or surrounding air.

Americium (Am): A silvery metal that is a man-made element whose isotopes Am-237 through Am-246 are all radioactive; trace quantities of americium are widely used in smoke detectors and as neutron sources in neutron moisture gauges.

Analysis approach: Selecting one of two primary approaches for FEMA: one is the hardware approach that lists individual hardware items and analyzes their possible failure modes, and the second is the functional approach that recognizes that every item is designed to perform a number of outputs within a system.

Anemometer: A rotating vane, swinging vane, or hot-wire device used to measure air velocity.

Anhydrous: A substance in which no water is present in the form of a hydrate or water of crystallization.

Anion: A negatively charged ion.

Annual survey: Survey conducted each year by the Bureau of Labor Statistics to produce national data on occupational injury and illness rates in various industries.

Annual summary: The occupational injury and illness totals for the year as reflected by OSHA Form 300 Log entries.

Anode: The positive electrode in an electrolytic cell.

Anosmia: Reduced sensitivity to odor detection.

Antidote: An agent that neutralizes or counteracts the effects of a poison.

Antimicrobial: Any agent that destroys microbial organisms.

Antiseptics: Substances applied to skin to reduce microbial flora such as alcohols, chlorine, and iodine.

Aperture: An opening through which radiation can pass.

Application program: The set of instructions that defines the specific intended tasks of robots and robot systems.

Approved: A method, procedure, equipment, or tool that has been determined to be satisfactory for a particular purpose.

Aqueous: A solution or suspension in which the solvent is water.

Area command: An organization established to oversee the management of multiple incidents that are each being handled by a separate incident command system organization or to oversee the management of a very large or evolving incident that has multiple incident management teams engaged.

Argon: A gas used as a laser medium that emits blue-green light.

Aromatic: Term applied to a group of hydrocarbons characterized by the presence of the benzene nucleus; a major series of unsaturated cyclic hydrocarbons whose carbon atoms are arranged in closed rings.

Asbestos: Fibrous magnesium silicate.

Asphyxiant: A chemical gas or vapor that can cause unconsciousness or death by suffocation.

Assessment: The evaluation and interpretation of measurements and other information to provide a basis for decision-making.

Assigned protection factor: A rating assigned to a respirator style by OSHA or NIOSH; this rating indicates the level of protection most workers can expect from the properly worn, maintained, and fitted respirator used under actual workplace conditions.

Assignment: A task given to a resource to perform within a given operational period that is based on operational objectives defined in the incident action plan.

Atom: The smallest particle of an element that can enter into a chemical reaction.

Atomic mass number: The total number of protons and neutrons in the nucleus of an atom.

Atomic mass unit: One unit is equal to 1/12th of the mass of a carbon-12 atom.

Atomic weight: The mass of an atom, expressed in atomic mass units.

Atmosphere-supplying respirator: A respirator that provides clean air from an uncontaminated source to the facepiece; examples include supplied-air (airline) respirators, self-contained breathing apparatus, and combination of supplied air/SCBA devices.

Attended continuous operation: The time when robots are performing (production) tasks at a speed no greater than slow speed through attended program execution.

Attended program verification: The time when a person within the restricted envelope (space) verifies the robot's programmed tasks at programmed speed.

Attenuated vaccine: A vaccine that has been weakened but is still required to be controlled as infectious by some regulatory programs.

Attenuation: The decrease in energy (or power) as a beam passes through an absorbing or scattering medium.

Autoclave: Device used to sterilize medical instruments and equipment by using steam under pressure.

Autoignition point: The lowest temperature at which a material will catch fire without the aid of a flame or spark.

Auto-ignition temperature: Temperature at which a material will self-ignite and maintain combustion without a fire source.

Automatic contour: A feature of a bed where the thigh section of the sleep surface articulates upward as the head section travels upward, thereby reducing the likelihood of patient/resident mattress from migrating toward the foot end of the bed.

Automatic conveyor and shuttle systems: Devices composed of various types of conveying systems linked together with various shuttle mechanisms for the prime purpose of conveying materials or parts to prepositioned and predetermined locations automatically.

Automatic guided vehicle system: Advanced material-handling or conveying systems that involve a driverless vehicle that follows a guide path.

Automatic mode: Robot state in which automatic operation can be initiated.

Automatic operation: Time during which robots are performing programmed tasks through unattended program execution.

Automatic sprinklers: System built in or added to a structure that automatically delivers water in case of fire.

Automatic storage and retrieval systems: Storage racks linked through automatically controlled conveyors and an automatic storage and retrieval machine or machines that ride on floor-mounted guide rails and power-driven wheels.

Awareness barrier: Physical and/or visual means that warn a person of an approaching or present hazard.

Awareness signal: A device that warns a person of an approaching or present hazard by means of audible sound or visible light.

Axis: The line about which a rotating body, such as a tool, turns.

B

Background radiation: The radiation in man's natural environment, including cosmic rays and radiation from naturally occurring radioactive elements.

Badging: Process of providing outside personnel with identification that gives them access to the designated facilities of the organization requesting assistance.

Barrier: A physical means of separating persons from the restricted envelope or space.

Base: Substance that (1) liberates hydroxyl ions when dissolved in water; (2) liberates negative ions of various kinds in any solvent; (3) receives a hydrogen ion from a strong acid to form a weaker acid; (4) gives up two electrons an acid, forming a covalent bond with the acid.

Basic human needs: Physical, safety, social, self-esteem, and self-actualization.

Basic life support: Noninvasive first-aid procedures and techniques utilized by all trained medical personnel, including first responders, to stabilize critically sick and injured people.

Beam: A collection of rays that may be parallel, convergent, or divergent.

Becquerel: The amount of a radioactive material that will undergo one decay (disintegration) per second.

Bed alarms: Alarms intended to notify caregivers of either an unwanted patient/resident egress or that the patient/resident is near the edge of the mattress.

Bed-rail extender: Detachable device intended to bridge the space between the head and foot bed rail.

Bed rails: Adjustable metal or rigid plastic bars that attach to the bed. Synonymous terms are side rails, bed side rails, and safety rails.

Beta particle: Particle emitted from a nucleus during radioactive decay that can be stopped by a sheet of metal or acrylic plastic, depending on the emitted energy level of a particular isotope.

Bioassay: An assessment of radioactive materials that may be present inside a person's body through analysis of the person's blood, urine, feces, or sweat.

Biocide: A substance that can kill living organisms.

Biodegradable: A substance with the ability to decompose or break down into natural components.

Biological half-life: The time required for one half of the amount of a substance, such as a radionuclide, to be expelled from the body by natural metabolic processes, not counting radioactive decay, once it has been taken in through inhalation, ingestion, or absorption.

Bioremediation: The management of microorganisms.

Bipolar: Forceps-shaped active electrode; current flows through the tissue from one tip to the other.

Block diagrams: These diagrams that illustrate that the operation, interrelationships, and interdependencies of the functions of a system are required to show the sequence and the series dependence or independence of functions and operations; block diagrams may be constructed in conjunction with or after defining the system and shall present the system breakdown of its major functions.

Blood-borne pathogens: Pathogenic microorganisms that are present in human blood and can cause disease in humans; these pathogens include, but are not limited to, hepatitis B virus and human immunodeficiency virus.

Blood-borne pathogens engineering controls: Sharps-disposal containers, self-sheathing needles, and safer medical devices, such as sharps with engineered sharps-injury protections and needleless systems that isolate or remove the blood-borne pathogens hazard from the workplace.

Blood-borne pathogens exposure incident: Means a specific eye, mouth, other mucous membrane, no intact skin, or parenteral contact with blood or other potentially infectious materials that result from the performance of an employee's duties.

Boiling point: The temperature at which a liquid changes to a vapor as expressed in degrees F at sea-level pressure.

Bolt ring: Closing device used to secure a cover to the body of an open head drum; this ring requires a bolt and nut to secure the closure.

Bonding: The interconnecting of two objects, such as a tank or cylinder, with clamps and wire as a safety practice to equalize the electrical potential between the objects and help prevent static sparks that could ignite flammable materials; dispensing/receiving a flammable liquid requires dissipating the static charge by bonding between containers.

Branch: Organizational level having functional or geographical responsibility for major aspects of incident operations, organizationally situated between the section chief and the division or group in the operations section.

Breakthrough time: The time from initial chemical contact to detection.

Buffer: An acid–base balancing or control reaction in which a pH of a solution is protected from major change when acids or bases are added to it.

Building-related illness: Diagnosable illnesses with identifiable symptoms that can be attributed to building contaminants.

Bung: A threaded closure located in the head or body of a drum.

Bureaucratic organization: Line organization with an established hierarchy.

Burn back: The distance a flame will travel from the ignition source back to the aerosol container.

C

Cache: A predetermined complement of tools, equipment, and/or supplies stored in a designated location, available for incident use.

Canister or cartridge: A container with a filter, sorbent, or catalyst, or a combination of these items, which removes specific contaminants from the air passed through the container.

Capture velocity: Term referring to air produced by a hood to capture outside contaminants.

Carbon dioxide: A heavy, colorless, nonflammable, relatively nontoxic gas produced by the combustion and decomposition of organic substances and as a by-product of many chemical processes.

Carbon monoxide: A colorless, odorless, toxic gas generated by the combustion of common fuels in the presence of insufficient air or where combustion is incomplete.

Carbonate: A compound formed by the reaction of carbonic acid with either a metal or an organic compound.

Carcinogen: Any substance that has been found to induce the formation of cancerous tissue in experimental animals.

Carpal tunnel syndrome: A common affliction caused by compression of the median nerve in the carpal tunnel, often associated with tingling, pain, or numbness in the thumb and first three fingers.

CAS: Chemical abstracts service.

CAS number: A number assigned to identify a chemical substance. Chemical abstracts service indexes information that appears in chemical abstracts, published by the American Chemical Society.

Catalyst: An element or compound that accelerates the rate of a chemical reaction but is neither changed nor consumed by it.

Catastrophic incident: Any natural or man-made incident, including terrorism, that results in extraordinary levels of mass casualties, damage, or disruption, severely affecting the population, infrastructure, environment, economy, national morale, and/or government functions.

Catastrophic loss: A loss of huge and extraordinary proportion.

Cathode: The negative electrode of an electrolytic cell.

Caustic material: That which is able to burn, corrode, dissolve, or eat away another substance.

Caustic substances: Strong alkalis, their solutions being corrosive to the skin and other tissues.

CBRN: Chemical, biological, radiological, or nuclear agent or substance.

Chemotherapy: Development and use of chemical compounds that are specific for the treatment of diseases.

Ceiling concentration: Maximum concentration of a toxic substance allowed at any time or during a specific sampling period.

Ceiling limit: Normally expressed as TLV and PEL, ceiling limit is the maximum allowable concentration to which an employee may be exposed in a given time period.

Ceiling maximum: Allowable exposure limit not to be exceeded for an airborne substance.

Ceiling value: Concentration that should not be exceeded during the working exposure; exposure should at no time exceed the ceiling value.

CFC: Chlorofluorocarbons, which are being phased out worldwide because of their detrimental effect on the ozone layer.

CFM: Cubic feet per minute, a unit measuring airflow when evacuating ventilation systems.

Colorimetry: An analytical method by which the amount of a compound in solution can be determined by measuring the strength of its color by either visual or photometric methods.

Chain of command: A series of command, control, executive, or management positions in hierarchical order of authority.

Chain reaction: A process that initiates its own repetition; in a fission chain reaction, a fissile nucleus absorbs a neutron and fissions, spontaneously releasing additional neutrons.

Characteristic waste: According to the EPA, hazardous waste can exhibit one of four characteristics: ignitability, reactivity, toxicity, or corrosivity.

Chemical disinfection: Use of formulated chemical solutions to treat and decontaminate infectious waste.

Chemical family: A group of compounds with related chemical and physical properties.

Chemical hygiene plan: A written plan that addresses job procedures, work equipment, protective clothing, and training necessary to protect employees from chemical and toxic hazards, required by OSHA under its laboratory safety standard.

Chemical name: Scientific designation of a chemical substance.

Chemical transportation emergency center: An organization that provides immediate information for members on what to do in case of spills, leaks, fires, or exposures.

Chief: The incident command system title for individuals responsible for management of functional sections: operations, planning, logistics, finance/administration, and intelligence/investigations, if established as a separate section.

Citizen Corps: A community-level program, administered by the department of homeland security, that brings government and private-sector groups together and coordinates the emergency preparedness and response activities of community members.

Chlorinated solvent: Organic solvent that contains chlorine atoms.

Chronic effect: Adverse effect on animals or humans in which symptoms develop slowly over a long period of time or recur frequently.

Chronic exposure: Exposure to a substance over a long period of time, possibly resulting in adverse health effects.

Class A fire: That which involves wood, paper, cloth, trash, or other ordinary materials.

Class B fire: That which involves grease, paint, or other flammable liquids.

Class C fire: That which involves live electrical or energized equipment.

Class D fire: That which involves flammable metals.

Class K fire: That which involves kitchen oils used for frying.

Clean Air Act: Public Law Pl 91-604, the EPA sets national ambient air quality standards, and enforcement/discharge permits are carried out by the states under implementation plans.

Clinical laboratory: Workplace where diagnostic or other screening procedures are performed on blood or other potentially infectious materials.

Cobalt (Co): Gray, hard, magnetic, and somewhat malleable metal, cobalt is relatively rare and generally obtained as a by-product of other metals, such as copper; the most common radioisotope is cobalt-60, used in radiography and medical applications.

Collective dose: The estimated dose for an area or region multiplied by the estimated population in that area or region.

Combustible: A term used to classify certain liquids that will burn on the basis of flash point; NFPA and DOT classify combustible liquids as having a flash point of 100°F (38°C) or higher.

Combustible liquids: OSHA defines combustible liquids as any liquid having a flash point at or above 100°F (38°C) but below 200°F (93.3°C).

Command: The act of directing, ordering, or controlling by virtue of explicit statutory, regulatory, or delegated authority.

Command staff: An incident command element consisting of the following functions: public information officer, safety officer, liaison officer, and other positions as required (all report to the incident commander).

Committed dose: A dose that accounts for continuing exposures expected to be received over a long period of time from radioactive materials that were deposited inside the body.

Common operating picture: An overview of an incident by all relevant parties that provides incident information, enabling the incident commander/unified command and any supporting agencies and organizations to make effective, consistent, and timely decisions.

Communications/dispatch center: An agency or interagency dispatch center, 911 call centers, emergency control or command dispatch centers, or any naming convention given to the facility and staff that handle emergency calls from the public and communication with emergency management/response personnel.

Compliance safety and health officer: An OSHA representative whose primary job is to conduct workplace inspections.

Comprehensive preparedness guide 101: FEMA document used to develop emergency plans for all-hazard emergency operations planning for state, territorial, local, and tribal governments.

Concentration: The ratio of the amount of a specific substance in a given volume or mass of solution to the mass or volume of solvent.

Concept plan or conplan: Any plan that describes the concept of operations for integrating and synchronizing federal capabilities to accomplish critical tasks and describes how federal capabilities will be integrated into and support regional, state, and local plans to meet the objectives described in the strategic plan.

Conductivity: The property of a circuit that permits the flow of an electrical current.

Consensus standards: A variety of standards developed according to a consensus of agreement among several organizations, stakeholders, or individuals.

Contaminated: The presence or the reasonably anticipated presence of blood or other potentially infectious materials on an item or surface.

Contaminated laundry: Laundry that has been soiled with blood or other potentially infectious materials or may contain sharps.

Contaminated sharps: Any contaminated object that can penetrate the skin including, but not limited to, needles, scalpels, broken glass, broken capillary tubes, and exposed ends of dental wires.

Contamination (radioactive): The deposit of unwanted radioactive material on the surfaces of structures, areas, objects, or people, where it may be external or internal.

Control bed rail: A bed rail that incorporates bed-function controls for patient/staff activation.

Control device: Any control hardware providing a means for human intervention in the control of a robot or robot system, such as an emergency-stop button, a start button, or a selector switch.

Controlling: Measuring performance of work by monitoring outcomes.

Controls program: The inherent set of control instructions that defines the capabilities, actions, and responses of the robot system not intended to be modified by the user or operator.

Coordinate: To advance systematically an analysis and exchange of information among principals who have or may have a need to know certain information to carry out specific incident-management responsibilities.

Coordinated straight-line motion: Control wherein the axes of the robot arrive at their respective end points simultaneously, giving a smooth appearance to the motion.

Corrective actions: Implementation of procedures that are based on lessons learned from actual incidents or during realistic exercises.

Corrosive: A substance that causes visible destruction or permanent change in human skin tissue at the site of contact.

Cosmic radiation: Radiation produced in space that heavily bombards the Earth.

Cost-benefit analysis: Evaluation of a situation that focuses on comparing expenditures with potential benefits but not necessarily a dollar-for-dollar comparison.

Counterterrorism security group: An interagency body convened on a regular basis to develop terrorism prevention policy and to coordinate threat response and law enforcement investigations associated with terrorism.

Credentialing: Authentication and verification of the certification and identity of designated incident managers and emergency responders.

Criteria standard: A standard against which performance can be measured.

Critical infrastructure: Systems, assets, and networks, whether physical or virtual, so vital to the United States that the incapacity or destruction of such systems and assets would have a debilitating impact on security, national economic security, national public health or safety, or any combination of these.

Critical mass: The minimum amount of fissile material that can achieve a self-sustaining nuclear chain reaction.

Criticality: A relative measure of the consequences of a failure mode and its frequency of occurrence.

Criticality (nuclear): A fission process where the neutron production rate equals the neutron loss rate to absorption or leakage.

Criticality analysis: A procedure by which each potential failure mode is ranked according to the combined influence of severity and probability of occurrence.

Cubic feet per minute: Measure of the volume of a substance flowing through air within a specified time period and used to measure air exchanged in ventilation systems.

Culture of trust: A culture where workers have a voice and choice in organizational matters.

Cumulative dose: The total dose resulting from repeated or continuous exposures of the same portion of the body, or of the whole body, to ionizing radiation.

Curie (Ci): Traditional measure of radioactivity, based on the observed decay rate of 1 g of radium.

Cutaneous radiation syndrome: Effects can be reddening and swelling of the exposed area, creating blisters, ulcers on the skin, hair loss, and severe pain.

D

Damper: Control that varies airflow through an air inlet, outlet, or duct.

Decay (radioactive): Disintegration of the nucleus of an unstable atom by the release of radiation.

Decay products: Isotopes or elements formed and the particles and high-energy electromagnetic radiation emitted by the nuclei of radionuclide during radioactive decay.

Decibel: A unit to express the relative intensity of a sound on a scale from 0 to 130 (average pain level); sound doubles every 10 decibels.

Decomposition: Breakdown of a chemical or substance into different parts or simpler compounds, which can occur due to heat, chemical reaction, or decay.

Decontamination: Process of removing contaminants from the body or a surface.

Defatting: Removal of natural oils from the skin by the use of a fat-dissolving solvent.

Demand respirator: An atmosphere-supplying respirator that admits breathing air to the facepiece only when a negative pressure is created inside the facepiece by inhalation.

Denaturant: A substance added to ethyl alcohol to prevent it being used for internal consumption.

Denier: A term used in the textile industry to designate the weight per unit length of a filament.

Density: The ratio of weight (mass) to volume of any substance, usually expressed as g/cm^3.

Density mass: Substance of mass per unit volume, usually compared to water, which has a density of 1.

Department operating center: An emergency operations center specific to a single department or agency.

Deposition density: The activity of a radionuclide per unit area of ground.

Dequervain's disease: The tendon sheath of both the long and the short abductor muscles of the thumb narrows, resulting in wrist deviation.

Dermatitis: Inflammation of the skin caused by defatting of the dermis.

Desiccant chemical: A substance that absorbs moisture.

Detection mechanism: The means or methods by which a failure can be discovered by an operator under normal system operation or can be discovered by the maintenance crew by some diagnostic action.

Deterministic effects: Effects that can be related directly to the radiation dose received, with severity increasing as the dose increases.

Deuterium: A nonradioactive isotope of the hydrogen atom that contains a neutron in its nucleus in addition to the one proton normally seen in hydrogen.

Device: Any control hardware such as an emergency-stop button, selector switch, control pendant, relay, solenoid valve, or sensor.

DFM: This abbreviation refers to a respirator filter cartridge suitable for use against dusts, fumes, or mist and is used in the new NIOSH regulation on respirator certification.

Dielectric material: An electrical insulator or in which an electric field can be sustained with a minimum dissipation of power.

Dilution ventilation: Process that uses an air-purification device to return exhaust to work-area air.

Directing: Providing the necessary guidance to others during job accomplishment.

Dirty bomb: A device designed to spread radioactive material by conventional explosives when the bomb detonates.

Disinfectant: An agent with the ability to kill at least 95% of targeted microorganisms.

Disaster: Any natural catastrophe, including any hurricane, tornado, storm, high water, wind-driven water, tidal wave, tsunami, earthquake, volcanic eruption, landslide, mudslide, snowstorm, and drought, or, regardless of cause, any fire, flood, or explosion.

Disaster recovery center: Facility established in a centralized location within or near the disaster area at which disaster victims apply for disaster aid.

Division: The partition of an incident into geographical areas of operation.

Doff: To take off or remove.

Don: To put on, in order to wear.

Dose (radiation): Radiation absorbed by person's body.

Dose coefficient: The factor used to convert radionuclide intake to dose.

Dose equivalent: A quantity used in radiation protection to place all radiation on a common scale for calculating tissue damage.

Dose quantity: Radiation absorbed per unit of mass by the body or by any portion of the body.

Dose rate: The radiation dose delivered per unit of time.

Dose reconstruction: A scientific study that estimates doses to people from releases of radioactivity or other pollutants.

Dose response: Relationship between the amount of a toxic or hazardous substance and the extent of illness or injury produced in humans.

Dosimeter: Small portable instrument (such as a film badge, thermoluminescent dosimeter, or pocket dosimeter) for measuring and recording the total accumulated dose of ionizing radiation a person receives.

Dosimetry: Assessment (by measurement or calculation) of radiation dose.

Drive powers: The energy source or sources for the robot actuators.

Drop test: A test required by DOT regulations for determination of the quality of a container or finished product.

Dry-bulb temperature: Temperature of air measured with a dry-bulb thermometer in a psycho-motor to measure relative humidity.

Dry pipe: Piping under pressure, and when head opens air is released and water flows into the system.

Dusts: Solid particles generated by handling, crushing, grinding, rapid impact, detonation, and decrepitating of organic or inorganic materials such as rock, ore, metal, coal, wood, and grain.

E

Effective dose: A dosimetry quantity useful for comparing the overall health effects of irradiation of the whole body; effective dose is used to compare the overall health detriments of different radionuclides in a given mix.

Effective half-life: The time required for the amount of a radionuclide deposited in a living organism to be diminished by 50% as a result of the combined action of radioactive decay and biologic elimination.

Elastomer: A term used to describe any high polymer having the essential properties of vulcanized natural rubber.

Electrochemistry: Chemistry concerned with the relationship between electrical forces and chemical reactions.

Electrode: A material used in an electrolytic cell to enable the current to enter or leave the solution.

Electrolysis: Decomposition of a chemical compound by means of an electric current.

Electron: A particle of negative electricity, electrons surround the nucleus of an atom, because of the attraction between their negative charge and the positive charge of the nucleus.

Electron volt: Unit of energy equivalent to the amount of energy gained by an electron when it passes from a point of low potential to a point one volt higher in potential.

Electrosurgery: Radio-frequency energy to produce cutting and coagulation in body tissues.

Electrosurgical unit: A device that produces radio-frequency energy for electrosurgery procedures.

Element: All isotopes of an atom that contain the same number of protons, or, in a nuclear facility, the fuel element is a metal rod containing the fissile material.

Embedded laser: A laser with an assigned class number higher than the inherent capability of the laser system in which it is incorporated.

Emergency: Any incident, whether natural or man-made, that requires responsive action to protect life or property.

Emergency-management-assistance compact: A congressionally ratified organization that provides form and structure to interstate mutual aid during emergency-event response.

Emergency-management committee: A preparedness entity established by an organization that has the responsibility for emergency-management-program oversight within the organization.

Emergency-management program: A program that implements the organization's mission, vision, management framework, and strategic goals and objectives related to emergencies and disasters.

Emergency manager: A person who has the day-to-day responsibility for emergency management programs and activities; roles include coordinating mitigation, preparedness, response, and recovery.

Emergency-operations center: Physical location at which the coordination of information and resources to support incident management and on-scene operations.

Emergency-operations plan: An ongoing plan maintained by various jurisdictional levels for responding to a wide variety of potential hazards, disasters, or emergency events.

Emergency public information: Information that is disseminated primarily in anticipation of an emergency or during an emergency, providing guidance or requiring actions.

Emergency response guide: A document that provides guidance on emergency response in a transportation incident involving a particular chemical.

Emergency situation: Any occurrence including equipment failure, rupture of containers, or failure of control equipment that results in an uncontrolled substantial release of a contaminant.

Emergency stop: The operation of a circuit using hardware-based components that override all other robotic controls.

Emergency support function: Refers to a group of capabilities of federal departments and agencies to provide the support, resources, program implementation, and services that are most likely to be needed to save lives, protect property, restore essential services and critical infrastructure, and help victims return to normal following a national incident.

Employee exposure: An exposure to a concentration of an airborne contaminant that would occur if the employees were not using respiratory protection.

Emulsion: Stable mixture of two or more liquids held in suspension by small percentages of substances called emulsifiers.

Enabling device: A manually operated device that permits motion when continuously activated, releasing the device stops robot motion and motion of associated equipment that may present a hazard.

Enclosed laser device: Any laser or laser system located within an enclosure that does not permit hazardous optical radiation emission from the enclosure.

End effector: An accessory device or tool specifically designed for attachment to the robot wrist or tool mounting plate to enable the robot to perform its intended task.

End-of-service-life indicator: A system that warns the respirator user of the approach of the end of adequate respiratory protection.

Endothermic: A term used to characterize a chemical reaction that requires absorption of heat from an external source.

Energy sources: Any electrical, mechanical, hydraulic, pneumatic, chemical, thermal, or other sources.

Engineering controls: The preferred method of controlling employee exposures in the workplace.

Enriched uranium: Any uranium in which the proportion of the isotope uranium-235 has been increased by removing uranium-238 mechanically.

Entrapment: An event in which a patient is caught, trapped, or entangled in the spaces in or about the bed rail, mattress, or hospital bed frame.

Envelope: Space or maximum volume of space encompassing the maximum designed movements of all robot parts, including the end effector, workpiece, and attachments.

Enzyme complex: Protein produced by living cells that starts up biochemical reactions.

Epidemiology: Study of the distribution and determinants of health-related states or events in specified populations and the application of this study to the control of health problems.

Ergonomics: A multidisciplinary activity that deals with interactions between workers and their total working environment plus stresses related to such environmental elements as atmosphere, heat, light, and sound, as well as tools and equipment in the workplace.

Escape-only respirator: A respirator intended to be used only for emergency exit.

Etiologic agent: Viable microorganism or its toxin that can cause human disease.

Evacuation: Organized, phased, and supervised withdrawal, dispersal, or removal of humans from dangerous or potentially dangerous areas.

Evaluating: Assessing the effectiveness for the purpose of improving.

Evaporation rate: The rate at which a material is converted to a vapor at a given temperature and pressure when compared to the evaporation rate of a given substance.

Excimer: A gas mixture used as the active medium in a family of lasers emitting ultraviolet light.

Exhaust ventilation: The removal of air from any space, usually by mechanical means.

Exothermic: A term used to characterize a chemical reaction that gives off heat as it proceeds.

Experience rating: Process of basing insurance or workers compensation premiums on the insured record of losses.

Explosion class 1: Flammable gas/vapor.

Explosion class 2: Combustible dust.

Explosion class 3: Ignitable fibers.

Explosion-proof can: An electrical apparatus designed so that the explosion of flammable gas or vapor inside an enclosure will not ignite flammable gas or vapor outside.

Exposure level: The level or concentration of a physical or chemical hazard to which an individual is exposed.

Explosive limit: The amount of vapor in the air that forms an explosive mixture.

Exposure limit: The concentration of a substance under which it is believed that nearly all workers may be repeatedly exposed day after day without adverse effects.

Exposure pathway: A route by which a radionuclide or other toxic material can enter the body.

Exposure (radiation): A measure of ionization in air caused by X-rays or gamma rays only; the unit of exposure most often used is roentgen.

Exposure rate: A measure of the ionization produced in air by X-rays or gamma rays per unit of time.

F

Face velocity: The average air velocity in the exhaust system measured at the opening of the hood or booth.

Fahrenheit: The temperature scale commonly used in the United States, with the freezing point of water 32°F and the boiling point 212°F at sea level.

Fail-safe interlock: An interlock where the failure of a single mechanical or electrical component of the interlock will cause the system to go into, or remain in, a safe mode.

Failure cause: The physical or chemical process, design defects, part misapplication, quality defects, or other processes that are the basic reason for failure or that initiate the physical process by which deterioration proceeds to failure.

Failure definition: This is a general statement of what constitutes a failure of the item in terms of performance parameters and allowable limits for each specified output.

Failure effect: Consequence(s) a failure mode has on the operation, function, or status of an item, failure effects are usually classified according to how the entire system is impacted.

Failure mode: The way by which failure is observed, failure mode describes the way the failure occurs and its impact on equipment operation.

Failure mode and effects analysis (FMEA): Process by which each potential failure mode in a system is analyzed to determine the results, or effects thereof, on the system and to classify each potential failure mode according to its severity.

Fallout (nuclear): Minute particles of radioactive debris that descend slowly from the atmosphere after a nuclear explosion.

First aid (OSHA): Any one-time treatment and subsequent observation of minor scratches, cuts, burns, and splinters that normally does not require medical care.

Film badges: A package of photographic film worn like a badge by persons working with or around radioactive material to measure exposure to ionizing radiation; the absorbed dose can be calculated from the degree of film darkening caused by the irradiation.

Filter: An air-purifying component used in respirators to remove solid or liquid aerosols from the inspired air.

Filtering facepiece (dust mask): A negative-pressure particulate respirator with a filter as an integral part of the facepiece or with the entire facepiece composed of the filtering medium.

Finance/administration section: The ICS functional area that addresses the financial, administrative, and legal/regulatory issues for the incident-management system.

Fireman pole: A pole secured (floor and ceiling mooring) next to the bed that acts as a support for the patient to get into and out of the bed.

First receiver: Employees at a hospital engaged in decontamination and treatment of victims who have been contaminated by a hazardous substance(s) during an emergency incident.

First report: A state-mandated worker compensation form used to report work-related injuries and illnesses.

First responder: Personnel who have responsibility to initially respond to emergencies, such as firefighters, police officers, highway patrol officers, lifeguards, forestry personnel, ambulance attendants, and other public service personnel; the first personnel trained to arrive on the scene of a hazardous or emergency situation.

First-responder operations level: Individuals who respond to releases or potential releases of hazardous substances as part of the initial response to the site for the purpose of protecting nearby persons, property, or the environment from the effects of the release; OSHA mandates these individuals must receive at least 8 hours of training or have sufficient experience to objectively demonstrate competency in specific critical areas.

Fission: Splitting of a nucleus into at least two nuclei that release a large amount of energy.

Fit factor: A quantitative estimate of the fit of a particular respirator to a specific individual, a fit factor typically estimates the ratio of the concentration of a substance in ambient air to its concentration inside the respirator when worn.

Fit test: The use of a protocol to qualitatively or quantitatively evaluate the fit of a respirator on an individual; fit testing can be qualitative or quantitative.

Flame arrestors: Mesh or perforated metal insert within a flammable storage can that protects its contents from external flame or ignition.

Flame extension: The distance a flame will travel from an aerosol container when exposed to an ignition source.

Flame retardant: Substances applied to or incorporated in a combustible material to reduce or eliminate its tendency to ignite when exposed a low-energy flame.

Flammable liquid: A liquid with a flash point below 100°F (37.8°C).

Flash back: A phenomenon characterized by vapor ignition and flame traveling back to the vapor source.

Flash point: The temperature at which an organic liquid evolves a high-enough concentration vapor at or near its surface to form an ignitable mixture with air.

Flocculation: The process to make solids in water increase in size by biological or chemical means, so that they may be separated from the water.

Fluorocarbon: Any of a broad group of organic compounds analogous to hydrocarbons, in which all or most of the hydrogen atoms have been replaced by fluorine.

Flux: Any material or substance that will reduce the melting or softening temperature of another material when added to it.

Fractionated exposure: An exposure to radiation that occurs in several small, acute exposures, rather than continuously, as in a chronic exposure.

Frequency: Electrical term indicating the number of wave cycles in a second, measured in units called hertz.

Fumes: Particulate matter consisting of the solid particles generated by condensation from the gaseous state, generally after violation from melted substances, and often accompanied by a chemical reaction such as oxidation.

Function: One of the five major activities in the incident command system—command, operations, planning, logistics, and finance/administration—the term is also used when describing the activity involved a planning function.

Functional approach: The functional approach is normally used when hardware items cannot be uniquely identified or when system complexity requires analysis from the top down.

Functional area: A major grouping of the similar tasks that agencies perform in carrying out incident-management activities.

Functional block diagrams: Diagrams that illustrate the operation and interrelationships between functional entities of a system as defined in engineering data and schematics.

Fungi: Organisms that lack chlorophyll and must receive food from decaying matter.

Fusion: A reaction in which at least one heavier, more stable nucleus is produced from two lighter, less stable nuclei.

G

Galvanizing: Application of a protective layer of zinc to a metal, chiefly steel, to prevent or inhibit corrosion.

Gamma rays: High-energy electromagnetic radiation emitted by certain radionuclides when their nuclei transition from a higher to a lower energy state, very similar to X-rays.

Gas: A state of matter in which a material has very low density and viscosity, gases expand and contract greatly in response to changes in temperature and pressure.

Gas discharge laser: A laser containing a gaseous lasing medium in a glass tube in which a constant flow of gas replenishes the molecules depleted by the electricity or chemicals used for excitation.

Gas/vapor sterilization waste treatment: A technique that uses gases or vaporized chemicals such as ethylene oxide and formaldehyde as sterilizing agents.

Gauge: Thickness of the steel used to manufacture a drum; the lower the gauge, the thicker the material.

Geiger counter: A radiation-detection and measuring instrument consisting of a gas-filled tube containing electrodes, between which there is an electrical voltage but no current flow; Geiger counters are the most commonly used portable radiation-detection instruments.

General staff: A group of incident management personnel organized according to function and reporting to the incident commander.

Generator: EPA term for any person, organization, or agency whose act or process produces medical waste or causes waste to become subject to regulation.

Genetic effects: Hereditary effects (mutations) that can be passed on through reproduction because of changes in sperm or ova.

Glacial: A term applied to a number of acids, which, in a highly pure state, have a freezing point slightly below room temperature.

Gram: A standard unit of mass (weight) equivalent to 1/453.49 pound.

Gravimetric: A term used by analytical chemists to denote methods of quantitative analysis that depend on the weight of the components in the sample.

Gray (Gy): A unit of measurement for the amount of energy absorbed in a material.

H

Half-life: The time any substance takes to decay by half of its original amount.

Hand antisepsis: Refers to either antiseptic hand wash or antiseptic hand rub.

Handgrips: Devices attached to either side of the bed to provide the patient/resident the ability to reposition themselves while in bed as well as an aid to enter and leave the bed.

Hand hygiene: A general term that applies to either hand washing, antiseptic hand-wash, antiseptic hand rub, or surgical hand antisepsis.

Hardware approach: The hardware approach is normally used when hardware items can be uniquely identified from schematics, drawings, and other engineering and design data. This approach is recommended for use in a part level-up approach, often referred to as the bottom-up approach.

Hazard: A potential or actual force, physical condition, or agent with the ability to cause human injury, illness, and/or death and significant damage to property, the environment, critical infrastructure, agriculture and business operations, and other types of harm or loss.

Hazard classes (DOT): The nine descriptive terms established by the United Nations committee of experts to categorize hazardous chemical, physical, and biological materials. Categories are flammable liquids, explosives, gaseous oxidizers, radioactive materials, corrosives, flammable solids, poisons, infectious substances, and dangerous substances.

Hazard control management: The practice of identifying, evaluating, and controlling hazards to prevent accidents, mitigate harm, prevent injury, and limit damage to property or the environment.

Hazard identification: A process to identify hazards and associated risk to persons, property, and structures and to improve protection from natural and human-caused hazards.

Hazard operability study: A structured means of evaluating a complex process to find problems associated with operability or safety of the process.

Hazard rating (NFPA): Classification system that uses a four-color diamond to communicate health, flammability, reactivity, and specific hazard information for a chemical substance; a numbering system that rates hazards from 0 (lowest) to 4 (highest).

Hazard vulnerability analysis (HVA): A systematic approach to identifying all potential hazards that may affect an organization and assessing the probability of occurrence and the consequences for each hazard; the organization creates a prioritized comparison of the hazard vulnerabilities.

Hazard vulnerability analysis (health-care): The identification of potential emergencies and direct and indirect effects these emergencies may have on the health-care organization's operations and the demand for its services.

Hazardous chemical: Any chemical that poses a physical or health hazard.

Hazardous material (DOT): A substance or material that has been determined by DOT to pose an unreasonable risk to health, safety, and property when transported in commerce; 49 CFR 171.8.

Hazardous motion: Any motion of machinery or equipment that could cause personal physical harm.

Hazardous substance: Any substance to which exposure may result in adverse effects on the health or safety of employees.

Hazardous waste (EPA): Any solid or combination of solid wastes, which, because of its physical, chemical, or infectious characteristics, may pose a hazard when not managed properly.

HAZCOM: The OSHA hazard communication standard (29 CFR 1910.1200).

HAZMAT: Hazardous material.

HAZWOPER: The OSHA standard on hazardous waste operations and emergency response; (29 CFR 1910.120) paragraph (q) of the standard covers employers whose employees are engaged in emergency response to hazardous-substance releases.

Health hazard: A chemical for which there is statistically significant evidence that acute or chronic health effects may occur in exposed individuals.

Health physics: A scientific field that focuses on protection of humans and the environment from radiation. Health physics uses physics, biology, chemistry, statistics, and electronic instrumentation to help protect individuals from any damaging effects of radiation.

Health-care coalition: A group of individual health-care organizations in a specified geographic area that agree to work together to enhance their response to emergencies or disasters, composed of relatively independent organizations that voluntarily coordinate their response, does not conduct command or control, and operates consistent with multi-agency coordination (MAC) system principles to support and facilitate the response of its participating organizations.

Hearing conservation: Preventing or minimizing noise-induced deafness through the use of hearing-protection devices, engineering methods, annual audiometric tests, and employee training.

Helium neon laser: A laser in which the active medium is a mixture of helium and neon, its wavelength is usually in the visible range.

High-efficiency particulate air filter: Any filter with at least 99.97% efficiency in the filtration of airborne particles 0.3 microns in diameter or greater.

High-impact mat: A mat placed next to the bed that absorbs the shock if the patient falls from the bed.

High-level radioactive waste: The radioactive material resulting from spent nuclear fuel reprocessing; this can include liquid waste directly produced in reprocessing or any solid material derived from the liquid wastes having a sufficient concentration of fission products.

Hospital decontamination zone: A zone that includes any areas where the type and quantity of hazardous substance are unknown and where contaminated victims, contaminated equipment, or contaminated waste may be present; this zone is sometimes called the warm zone, contamination-reduction zone, yellow zone, or limited-access zone.

Hospital incident command system: Optional ICS tailored specifically for use by hospitals and designed to function in conjunction with other common command systems used by emergency response organizations.

Hospital postdecontamination zone: The hospital postdecontamination zone is an area considered uncontaminated, sometimes called the cold zone or clean area.

Hospital incident command system: A generic crisis management plan expressly for comprehensive medical facilities that is modeled closely after the fire service incident-command system.

Hot spot: Any place where the level of radioactive contamination is considerably greater than the area around it.

HSPD 5: Homeland security presidential directive that addresses the management of domestic incidents, including a national response plan to integrate all federal government domestic prevention, preparedness, response, and recovery plans into a single all-discipline, all-hazards plan; hospitals should be national incident management system compliant and develop an all-hazards approach to emergency management.

HSPD 6: Homeland security directive that addresses the integration and use of screening information to protect against terrorism and uses information as appropriate and to the full extent permitted by law to support (1) federal, state, local, territorial, tribal, foreign government, and private sector screening processes and (2) diplomatic, military, intelligence, law enforcement, immigration, visa, and protective processes.

HSPD 7: Homeland security presidential directive that addresses critical infrastructure, identification, prioritization, and protection.

HSPD 8: Homeland security presidential directive that addresses national preparedness and calls for a national preparedness goal that establishes measurable priorities and targets and an approach to developing needed capabilities; it clearly states that hospital emergency medical facilities are considered emergency response providers as defined by the Department of Homeland Security Act.

HSPD 9: Homeland security presidential directive that addresses the defense of US agriculture and food; the directive establishes a national policy to defend the agriculture and food system against terrorist attacks, major disasters, and other emergencies.

HSPD 12: Homeland security presidential directive that addresses common identification standard for federal employees and contractors; directive was issued in response to the fears of bioterrorism following the anthrax attacks of 2001 and to address the threat of pandemic influenza and severe acute respiratory syndrome.

HSPD 20: Homeland security presidential directive that addresses the national continuity policy that establishes a comprehensive national policy on the continuity of federal government structures and operations and also creates the position of a single national continuity coordinator responsible for coordinating the development and implementation of federal continuity policies.

HSPD 21: Homeland security presidential directive that addresses the national strategy for public health and medical preparedness.

Humidity (relative): The ratio of the amount of water vapor present in air at a given temperature to the maximum that can be held by air at that temperature.

Hydrocarbon: Any compound composed of carbon and hydrogen.

Hydrophilic: A term that refers to substances that tend to absorb and retain water.

Hydrophobic: A term that describes substances that repel water.

Hygroscopic: A term used to describe solid or liquid materials that pick up and retain water vapor from the air.

Hypersensitivity diseases: Diseases characterized by allergic responses to animal antigens, often associated with indoor air quality conditions such as asthma and rhinitis.

I

Ignitable solid, liquid, or compressed gas: Must have a flash point less than 140°F.

Ignition temperature: Lowest temperature at which a substance can catch fire and continue to burn.

Illumination: The amount of light a surface receives per unit area, expressed in lumens per square foot, or foot candles.

Immediate dangerous to life or health: An atmospheric concentration of any toxic, corrosive, or asphyxiate substance that poses an immediate threat to life or would interfere with an individual's ability to escape from a dangerous atmosphere.

Immiscible: A term used to describe substances of the same phase that cannot be uniformly mixed or blended.

Incident: An actual or impending unplanned event with hazard impact, either human-caused or by natural phenomena, that requires action by emergency personnel to prevent or minimize loss of life or damage to property and/or natural resources.

Incident action plan: An oral or written plan containing general objectives reflecting the overall strategy for managing an incident; it may include the identification of operational resources and assignments.

Incident command system: A flexible organizational structure that provides a basic expandable system developed by the fire services to mitigate any size emergency situation.

Incident commander: The individual responsible for all incident activities, including the development of strategies and tactics and the ordering and the release of resources.

Incident management: Refers to how incidents are managed across all homeland security activities, including prevention, protection, and response and recovery.

Incident management team: The incident commander and appropriate command and general staff personnel assigned to an incident.

Incident objectives: Statements of guidance and direction necessary for selecting appropriate strategies and the tactical direction of resources, incident objectives are based on realistic expectations of what can be accomplished when allocated resources have been effectively deployed.

Incompatible: The term used to indicate that one material cannot be mixed with another without the possibility of a dangerous reaction.

Indicator: A measurement used to evaluate program effectiveness within an organization.

Indoor air quality: The study, evaluation, and control of indoor air quality related to temperature, humidity, and building contaminants.

Industrial equipment: Physical apparatus used to perform industrial tasks, such as welders, conveyors, machine tools, fork trucks, turntables, positioning tables, or robots.

Industrial robot: A reprogrammable, multifunctional manipulator designed to move material, parts, tools, or specialized devices through variable programmed motions for the performance of a variety of tasks.

Industrial robot system: A system that includes industrial robots, the end effectors, and the devices and sensors required for the robots to be taught or programmed, or for the robots to perform the intended automatic operations, as well as the communication interfaces required for interlocking, sequencing, or monitoring the robots.

Inert: Having little or no chemical affinity or activity.

Infectious waste: Waste containing pathogens that can cause an infectious disease in humans.

Inhalation: Breathing of an airborne substance, which may be in the form of a gas, vapor, fume, mist, or dust, into the body.

Inhibitor: A substance that is added to another substance to prevent or slow down an unwanted reaction or change.

Innocuous: Harmless.

Inorganic: This term refers to a major and the oldest branch of chemistry; it is concerned with substances that do not contain carbon.

Iodine: A nonmetallic solid element, there are both radioactive and nonradioactive isotopes of iodine.

Ion: An atom, group, or molecule that has either lost one or more electrons or gained one or more electrons.

Ionization: The process of adding one or more electrons to, or removing one or more electrons from, atoms or molecules.

Ionizing radiation: Any radiation capable of displacing electrons from atoms, thereby producing ions, and high doses of ionizing radiation may produce severe skin or tissue damage.

Interior structural firefighting: The physical activity of fire suppression, rescue, or both, inside of buildings or enclosed structures that are involved in a fire situation beyond the incipient stage.

Interlock: An arrangement whereby the operation of one control or mechanism brings about or prevents the operation of another.

Internal exposure: Exposure to radioactive material taken into the body.

Interoperability: The ability of emergency management/response personnel to interact and work well together; in the context of technology, interoperability also refers to having an emergency communications system that is the same or is linked to the same system that a jurisdiction uses for nonemergency procedures and that effectively interfaces with national standards as they are developed.

Intra-beam viewing: The viewing condition whereby the eye is exposed to all or part of a direct laser beam or a specular reflection.

Irradiation: Exposure to radiation.

Irradiation sterilization: The use of ionizing radiation for the treatment of infectious waste.

Isomer: One of two or more compounds having the same molecular weight and formula, but often having quite different properties and somewhat different structures.

Isotonic: Having the same osmotic pressure as the fluid phase of a cell or tissue.

Isotope: A nuclide of an element having the same number of protons but a different number of neutrons; any of two or more forms of an element in which the weights differ by one or more mass units due to a variation in the number of neutrons in the nuclei.

J

Job hazard analysis: The breaking down of methods, tasks, or procedures into components to determine hazards.

Joint field office: The primary federal incident management field structure for support; the JFO uses an incident command system structure but does not manage on-scene operations.

Joint information center: A center established to coordinate the public information activities for a large incident.

Joint operations center: An interagency command post established by the FBI to manage terrorist threats or incidents and investigative and intelligence activities.

Joint motion: A method for coordinating the movement of the joints such that all joints arrive at the desired location simultaneously.

Joule: Unit of energy used to describe a single pulsed output of a laser; it is equal to 1 watt-second or 0.239 calories.

Jurisdiction: A political subdivision with the responsibility for ensuring public safety, health, and welfare within its legal authorities and geographic boundaries.

K

Ketone: A class of unsaturated and reactive compounds whose formula is characterized by a carbonyl group to which two organic groups are attached.

Kilogram: About 2.2 pounds.

Kiloton: The energy of an explosion that is equivalent to an explosion of 1000 tons of TNT.

Kinetic energy: The energy that a particle or an object possesses due to its motion or vibration.

L

Lab pack: Generally refers to any small container of hazardous waste in an overpacked drum. Not restricted to laboratory wastes.

Lacquer: A type of organic coating in which rapid drying is effected by evaporation of solvents.

Laser: A term for light amplification by stimulated emission of radiation; a laser is a cavity with mirrors at the ends, filled with materials such as crystal, glass, liquid, gas, or dye. It produces an intense beam of light with the unique properties of coherency, collimation, and monochromaticity.

Laser medium: Material used to emit the laser light and for which the laser is named.

Laser safety officer: Person with authority to monitor and enforce measures to control laser hazards and effect the knowledgeable evaluation and control of laser hazards.

Laser system: An assembly of electrical, mechanical, and optical components of a laser, under federal standard it also includes the power supply.

Latent period: Time between exposure to a toxic material and the appearance of a resultant health effect.

Leading: A person who creates an atmosphere and purpose that encourage people to succeed and achieve.

Leak test: A test performed to detect leakage of a radiation source.

Lens: A curved piece of optically transparent material that, depending on its shape, is used to either converge or diverge light.

Level of analysis: The level of analysis applies to the system hardware or functional level at which failures are postulated.

Liaison officer: A member of the command staff responsible for coordinating with representatives from cooperating and assisting agencies or organizations.

Light-emitting diode: A semiconductor diode that converts electric energy efficiently into spontaneous and noncoherent electromagnetic radiation at visible and near-infrared wavelengths.

Limiting aperture: Maximum circular area over which radiance and radiant exposure can be averaged when determining safety hazards.

Limiting device: A device that restricts the maximum envelope (space) by stopping or causing to stop all robot motion and is independent of the control program and the application programs.

Line organization: An organization with a chain-of-command hierarchy.

Liquefied petroleum: A gas usually comprised of propane and some butane, created as a by-product of petroleum refining.

Liquid crystal display: A constantly operating display that consists of segments of a liquid crystal whose reflectivity varies according to the voltage applied to the unit.

Local exhaust ventilation: A ventilation system that captures/removes contaminants at the point produced before they escape into the work area.

Long-term recovery: A process of recovery that may continue for a number of months or years, depending on the severity and extent of the damage sustained.

Loose-fitting facepiece: A respiratory-inlet covering that is designed to form a partial seal with the face.

Loss ratio: A fraction calculated by dividing losses of an organization and the amount of insurance premiums paid.

Lost workdays: The number of workdays an employee is away from work beyond the day of injury or onset of illness.

Low-level waste: Radioactively contaminated industrial or research waste such as paper, rags, plastic bags, medical waste, and water-treatment residues.

Lower explosive limit: The lowest concentration of a substance that will produce a fire or flash when an ignition source is present, expressed as a percent of vapor or gas in the air by volume.

Lumbar: The section of the lower vertebral column immediately above the sacrum, located in the small of the back and consisting of five large lumbar vertebrae.

M

Management by exception: A manager makes a decision by reviewing key information and not all available information on a subject.

Management by objective: A management theory where a manager and subordinates agree on a predetermined course of action or objective.

Management by objectives: ICS concept that relates to a proactive management activity that involves a four-step process to achieve the incident goal.

Manometer: A device that measures pressure differences in inches of a water gauge.

Mass: The amount of material substance present in a body, irrespective of gravity.

Mass casualty incident: An incident that generates a sufficiently large number of casualties whereby the available health-care resources, or their management systems, are severely challenged or unable to meet the health-care needs of the affected population.

Mass effect incident: An incident that primarily affects the ability of an organization to continue its normal operations.

Mass spectroscopy: Process that identifies compounds by breaking them up into all combinations of ions and measuring mass-to-charge ratios at detector.

Material safety data sheet: A document that contains descriptive information on hazardous chemicals under OSHA hazard communication standard; data sheets also provide precautionary information, safe handling procedures, and emergency first-aid procedures.

Maximum permissible exposure: The level of laser radiation to which a person may be exposed without hazardous effect or adverse biological changes in the eye or skin.

Measure: Term used in the quality field for the collection of quantifiable data and information about performance, production, and goal accomplishment.

Measures of effectiveness: Defined criteria for determining whether satisfactory progress is being accomplished toward achieving the incident objectives.

Medical surge: The ability to provide adequate medical evaluation and care in events that severely challenge or exceed the normal medical infrastructure of an affected community.

Medical waste: Any solid waste generated in the diagnosis, treatment, or immunization of humans or animals.

Megaton (Mt): The energy of an explosion that is equivalent to an explosion of 1 million tons of TNT. One megaton is equal to a quintillion (10^{18}) calories. *See also* kiloton.

Milligrams per cubic meter: Unit used to measure air concentration of dust, gas, mist, and fume.

Memorandum of agreement: A conditional agreement between two or more parties, one party's action depends on the other party's action.

Memorandum of understanding: A formal agreement documenting the commitment of two or more parties to an agreed undertaking.

Microbe: Minute organism, including bacteria, protozoa, and fungi, which is capable of causing disease.

Micron: A unit of length in the metric system equivalent to one-millionth of a meter.

Mitigation: Activities designed to reduce or eliminate risks to persons or property or to lessen the actual or potential effects or consequences of a hazard.

Mobile robots: Freely moving, automatic, programmable industrial robots.

Mobilization: Process and procedures used by all organizations for activating, assembling, and transporting all resources that have been requested to respond to or support a disaster incident.

Molecular weight: The total obtained by adding together the weights of all the atoms present in a molecule.

Molecule: A combination of two or more atoms that are chemically bonded. A molecule is the smallest unit of a compound that can exist by itself and retain all of its chemical properties.

Mutagen: A substance or agent capable of changing the genetic material of a living cell.

Multiagency coordination (MAC): A group of administrators/executives, or their appointed representatives, who are authorized to commit agency resources and funds and can provide coordinated decision-making and resource allocation among cooperating agencies.

Muting: The deactivation of a presence-sensing safeguarding device during a portion of the robot cycle.

Mutual aid agreement: Written instrument between agencies and/or jurisdictions in which they agree to assist one another upon request, by furnishing personnel, equipment, supplies, and/or expertise in a specified manner.

N

Naphtha: Any of several liquid mixtures of hydrocarbons of specific boiling and distillation ranges derived from either petroleum or coal tar.

Narcosis: Stupor or unconsciousness caused by exposure to a chemical.

National disaster medical system: A federally coordinated system that augments the nation's medical-response capability.

National incident management system (NIMS): A system that provides a proactive approach guiding government agencies at all levels, the private sector, and nongovernmental organizations to work seamlessly to prepare for, prevent, respond to, recover from, and mitigate the effects of incidents, regardless of cause, size, location, or complexity, to reduce the loss of life or property and harm to the environment.

National joint terrorism task force: Entity responsible for enhancing communications, coordination, and cooperation among agencies representing the intelligence, law enforcement, defense, diplomatic, public safety, and homeland security.

National response framework: Policy that guides how the nation conducts all-hazards response; framework documents the key response principles, roles, and structures that organize national response.

National Security Council: Advises the president on national strategic and policy during large-scale incidents.

Natural gas: A combustible gas composed largely of methane and other hydrocarbons obtained from natural earth fissures.

Necrosis: Death of plant or animal cells.

Needleless systems: Devices that do not use needles for the collection of bodily fluids or withdrawal of body fluids after initial venous or arterial access is established.

Negative pressure: A condition caused when less air is supplied to a space than is exhausted from the space. The air pressure in the space is less than that in surrounding areas.

Negligence: Failure to do what reasonable and prudent persons would do under similar or existing circumstances.

Neutralization: The reaction between equivalent amounts of an acid and a base to form a salt.

Neutron: A small atomic particle possessing no electrical charge, typically found within an atom's nucleus.

Nitrogen oxide: Compound produced by combustion.

NMR: Nuclear magnetic resonance.

Nomenclature: Names of chemical substances and the system used for assigning them.

Nonionizing radiation: Radiation that has lower energy levels and longer wavelengths than ionizing radiation; examples include radio waves, microwaves, visible light, and infrared.

Nonstochastic effects: Effects that can be related directly to the radiation dose received.

Nucleus: Central part of an atom that contains protons and neutrons.

Nuclide: A general term applicable to all atomic forms of an element.

Numerically controlled machine tools: Tools operated by a series of coded instructions composed of numbers, letters of the alphabet, and other symbols.

O

Occupational illness: Illness caused by environmental exposure during employment.

Occurrence: Incident classified as major or minor, which results from apparent or foreseen causal factors.

Odor threshold: Minimum concentration of a substance at which most people can detect and identify its characteristic odor.

Operating envelope: The portion of the restricted envelope (space) that is actually used by the robot while performing its programmed motions.

Operations plan: A plan developed by and for each federal department or agency describing detailed resource, personnel, and asset allocations necessary to support the concept of operations detailed in the concept plan.

Operator: The person designated to start, monitor, and stop the intended productive operation of a robot or robot system.

Optical cavity (resonator): Space between the laser mirrors where lasing action occurs.

Optical density: Logarithmic expression of the attenuation afforded by a filter.

Optical fiber: A filament of quartz or other optical material capable of transmitting light along its length by multiple internal reflections and emitting it at the end.

Order of magnitude: A term used in science to indicate a range of values representing numbers, dimensions, or distances that starts at any given value and ends at 10 times that value.

Organic: Any compound containing the element carbon; it also describes substances derived from living organisms.

Organizing: Arranging work or tasks to be performed in the most efficient manner.

Outcomes: Results reached due to performance or nonperformance of a task, job, or process.

Output power: Energy per second measured in watts emitted from the laser in the form of coherent light.

Overt culture: The formal, expected, published, visible, or anticipated culture of an organization.

Oxidant: An oxygen-containing substance that reacts chemically to produce a new substance.

Oxidation: Reaction in which electrons are transferred from one atom to another either in the uncombined state or within a molecule.

Oxidizers: Materials that may cause the ignition of a combustible material without the aid of an external ignition source.

Oxygen-deficient atmosphere: An atmosphere with oxygen content below 19.5% by volume.

Ozone: Reactive oxidant that contains three atoms of oxygen.

P

Parenteral: Piercing mucous membranes or the skin barrier through such events as needles ticks, human bites, cuts, and abrasions.

Particulates: Fine solid or liquid particles found in air and other emissions.

Parts per million: A unit for measuring the concentration of a gas or vapor in contaminated air, used to indicate the concentration of a particular substance in a liquid or solid.

Pasteurization: Heat treatment of liquid or semiliquid food products for the purpose of killing or inactivating disease-causing bacteria.

Pathways: The routes by which people are exposed to radiation or other contaminants. The three basic pathways are inhalation, ingestion, and direct external exposure.

Patient assessment: An assessment that provides ongoing information necessary to develop a care plan and provide the appropriate care and services for each patient.

PEL: An acronym for permissible exposure limit, PEL is the OSHA limit for employee exposure to chemicals (29 CFR 1910.1000), based on a TWA of hours for a 40-hour work week.

Pendant: Any portable control device, including teach pendants, that permits an operator to control the robot from within the restricted envelope space of a robot.

Pendant control: A means used by either the patient or the operator to control the drives that activate various bed functions and are attached to the bed by a cord.

Penetrating radiation: Any radiation that can penetrate the skin and reach internal organs and tissues.

Periodic law: States that the arrangement of electrons in the atoms of any given chemical element and the properties determined by this arrangement are closely related to the atomic number of that element.

Periodic table: A systematic classification of the chemical elements based on the periodic law.

Permeation rate: An invisible process by which a hazardous chemical moves through a protective material.

Persistent activity: The prolonged or extended antimicrobial activity that prevents or inhibits the proliferation or survival of microorganisms after application of an antimicrobial product.

pH: A scale indicating the acidity or alkalinity of aqueous solutions.

Physical hazard of a chemical: A chemical validated as being or having one of the following characteristics: combustible liquid, compressed gas, explosive, flammable, organic peroxide, oxidizing qualities, pyrophoric, unstable, or water-reactive.

Planning: Actions taken to predetermine the best course of action.

Planning section: Functional area is responsible for the collection, evaluation, and dissemination of operational information related to the incident and for the preparation and documentation of the incident action plan and its support.

Plume: Smoke from the use of electrosurgery, lasers, and aerosols.

Poison: Solid or liquid substance that is known to be toxic to humans.

Polychlorinated biphenyls: A pathogenic and teratogenic industrial compound used as a heat transfer agent, they accumulate in human or animal tissue.

Polymerization: A chemical reaction in which one or more small molecules combine to form larger molecules.

Polyvinyl chloride: A member of the family of vinyl resins.

Positive pressure: A respirator in which the pressure inside the respiratory inlet covering exceeds the ambient air pressure outside the respirator.

Powered air-purifying respirator: An air-purifying respirator that uses a blower to force the ambient air through air-purifying elements to the inlet covering.

Preaction: When the main water-control valve is opened by an actuating device.

Prefilter: A filter used in conjunction with a cartridge on an air-purifying respirator.

Preparedness: The range of deliberate, critical tasks and activities necessary to build, sustain, and improve the capability to protect against, respond to, and recover from hazard impacts.

Presence-sensing safeguarding device: A device designed, constructed, and installed to create a sensing field or area to detect an intrusion into the field or area by personnel, robots, or other objects.

Pressure demand respirator: A positive-pressure, atmosphere-supplying respirator that admits breathing air to the facepiece when the positive pressure is reduced inside the facepiece by inhalation.

Prevention: Actions to avoid an incident or to intervene to stop an incident from occurring.

Process: Method of interrelating steps, events, and mechanisms to accomplish an action or goal.

Processes: Systems of operations that incorporate standardized procedures, methodologies, and functions necessary to effectively and efficiently accomplish objectives.

Protocol: A set of established guidelines for actions that may be designated by individuals, teams, functions, or capabilities under various specified conditions.

Proton: Basic unit of mass that is a constituent of the nucleus of all elements, the number present being the atomic number of a given element.

Pulse duration: The "on" time of a pulsed laser, it may be measured in terms of milliseconds, microseconds, or nanoseconds.

Pyrolysis: A chemical change brought about by heat alone.

Pyrophoric: A chemical that will ignite spontaneously in air at a temperature of 130°F or below.

Q

Qualitative analysis: Examination of a sample of a material to determine the kinds of substances present and to identify each constituent.

Qualitative fit test: A pass/fail fit test to assess the adequacy of respiratory fit that relies on the individual's response to the test agent.

Quantitative fit test: An assessment of the adequacy of respirator fit by numerically measuring the amount of leakage into the respirator.

Quaternary ammonium compounds: Chemical substances used to disinfect or sanitize by rupturing the cell walls of microorganisms.

R

Rad (radiation absorbed dose): A basic unit of absorbed radiation dose.

Radiation warning symbol: A symbol prescribed by OSHA, it is a magenta on a yellow background, displayed where certain quantities of radioactive materials are present or where certain doses of radiation could be received.

Radioactivity: The spontaneous decay or disintegration of an unstable atomic nucleus, usually accompanied by the emission of ionizing radiation.

Radioassay: A test to determine the amounts of radioactive materials through the detection of ionizing radiation.

Radiography: Medical use of radiant energy (such as X-rays and gamma rays) to image body systems.

Radioisotope: Isotopes of an element that have an unstable nucleus, commonly used in science, industry, and medicine.

Reactivity: Susceptibility of a substance to undergo chemical reaction and change that could result in an explosion or fire.

Reagent: Any chemical compound used in laboratory analyses to detect and identify specific constituents of the material being examined.

Recommend exposure limit: A NIOSH chemical-exposure-limit recommendation.

Recovery: The phase of comprehensive emergency management that encompasses activities and programs implemented during and after response that are designed to return the entity to its usual state or a new normal operating status.

Reflection: Return of radiant energy (incident light) by a surface, with no change in wavelength.

Refraction: Change of direction of propagation of any wave, such as an electromagnetic wave, when it passes from one medium to another in which the wave velocity is different.

Relative humidity: The ratio of the quantity of water vapor present in air to the quantity that would saturate the air at any specific temperature.

Relative risk: The ratio between the risks for disease in an irradiated population to the risk in an unexposed population.

Release zone: An area in and immediately surrounding a hazardous-substance release.

Reliability block diagrams: Diagrams that define the series dependence, or independence, of all functions of a system or functional group for each life-cycle event.

Relief valve: A valve designed to release excess pressure within a system without damaging the system.

Rem: Roentgen equivalent man, the unit of dose of any ionizing radiation that produces the same biological effect on human tissue as 1 roentgen of X-rays.

Resiliency: The ability of an individual or organization to quickly recover from change or misfortune.

Resin: Naturally occurring water-insoluble mixtures of carboxylic acids, essential oils, and other substances formed in numerous varieties of trees and shrubs.

Resonator: Mirrors (or reflectors) making up the laser cavity, including the laser rod or tube. The mirrors reflect light back and forth to build up amplification.

Resource Conservation and Recovery Act: Legislation used by the EPA to regulate waste materials, including hazardous wastes from generation through final disposal.

Resource management: A system for identifying available resources at all jurisdictional levels to enable timely and unimpeded access to resources needed to prepare for, respond to, or recover from an incident.

Respiratory inlet covering: The portion of a respirator that forms the protective barrier between the user's respiratory tract and an air-purifying device or breathing air source.

Response: Activities that address the direct effects of an incident, response includes immediate actions to save lives, protect property, and meet basic human needs.

Reversible: A chemical reaction that can proceed first to the right and then to the left when the conditions change.

Reynaud's syndrome: A condition where the blood vessels in the hand constrict due to cold temperature, vibration, emotion, or unknown causes.

Right to know: Phrase that relates to an employee's right to know about the nature and hazards of agents used in the workplace and/or to the right of communities.

Risk: The probability of injury, illness, disease, loss, or death under specific circumstances.

Risk assessment: An evaluation of the risk to human health or the environment by hazards, risk assessments can look at either existing hazards or potential hazards.

Roentgen: A unit of exposure to X-rays or gamma rays.

Roentgen equivalent man: A unit of equivalent dose that relates the absorbed dose in human tissue to the effective biological damage of the radiation.

S

Safeguard: A barrier guard, device, or safety procedure designed for the protection of personnel.

Safety: Human actions to control, reduce, or prevent accidental loss.

Safety belt: A belt worn to prevent falls when working in high places; a belt used to secure passengers in vehicles or airplanes.

Safety can: An approved container of not more than 5-gallon capacity with a spring-closing lid and a spout cover designed to safely relieve internal pressure when exposed to fire.

Safety hat: A hard hat worn to protect a worker from head injuries, flying particles, and electric shock.

Safety procedure: An instruction designed for the protection of personnel.

Salt: One of the products resulting from a reaction between an acid and a base.

Sanitize: To destroy common microorganisms on a surface to a safe level.

Scanning laser: A laser having a time-varying direction, origin, or pattern of propagation with respect to a stationary frame of reference.

Section: The organizational level having responsibility for a major functional area of incident management such as operations, planning, logistics, finance/administration, and intelligence.

Self-contained breathing apparatus (SCBA): A respirator that provides fresh air to the facepiece from a compressed air tank.

Semiconductor: Laser-type of laser that produces its output from semiconductor materials.

Sensitivity: The ability of an analytical method to detect small concentrations of radioactive material.

Sensitizer: A substance that may cause no reaction in a person during initial exposure, but will cause an allergic response upon further exposure.

Sensor: A device that responds to physical stimuli such as heat, light, sound, pressure, magnetism, or motion.

Serious injury: An injury classification that includes disabling work injuries and injuries such as eye injuries, fractures, hospitalization for observation, loss of consciousness, and any other injury that requires medical treatment by a physician.

Severity: Consequences of a failure as a result of a particular failure mode.

Severity classification: A classification assigned to provide a qualitative measure of the worst potential consequences resulting from design error or item failure.

Service: To adjust, repair, maintain, and make fit for use.

Service life: The period of time that a respirator, filter or sorbent, or other respiratory equipment provide adequate protection to the wearer.

Service robots: Machines that extend human capabilities.

Sharps: Objects that can penetrate the skin, such as needles, scalpels, and lancets.

Shielding: Material between a radiation source and a potentially exposed person that reduces exposure.

Short-term exposure limit: An OSHA measurement of the maximum concentration for a continuous 15-minute exposure period.

Short-term recover: A process of recovery that is immediate and overlaps emergency-response actions.

Sick building syndrome: A situation where building occupants experience acute health or discomfort that appears to be linked to time spent in the building but for which no specific illness or cause can be determined.

Sievert: A unit used to derive a quantity called dose equivalent as it relates to the absorbed dose in human tissue to the effective biological damage of the radiation.

Single failure: Point of failure of an item that would result in failure of the system and is not compensated by redundancy or alternative operational procedure.

Situation report: Document that contains confirmed or verified information and explicit details (who, what, where, and how) relating to an incident.

Situational awareness: The ability to identify, process, and comprehend the critical elements of information about an incident.

Sludge: A solid material that collects as the result of air- or water-treatment processes.

Solubility: The percentage of a material (by weight) that will dissolve in water at a specified temperature.

Solvent: A substance that dissolves or disperses another substance.

Somatic effects: The effects of radiation that is limited to the exposed person.

Source: Laser or laser-illuminated reflecting surface.

Span of control: The number of individuals a supervisor is responsible for, usually expressed as the ratio of supervisors to individuals.

Specific gravity: The weight of a material compared to the weight of an equal volume of water.

Spectrum: A range of frequencies within which radiation has some specified characteristics, such as audiofrequency spectrum, ultraviolet spectrum, and radio spectrum.

Staging area: Any location in which personnel, supplies, and equipment can be temporarily housed or parked while awaiting operational assignment.

Standard industrial classification: A classification developed by the office of management and budget, used to assign each establishment an industry code that is determined by the product manufactured or service provided.

Standard operating procedure: Complete reference document or an operations manual that provides the purpose, authorities, duration, and details for the preferred method of performing a single function or a number of interrelated functions in a uniform manner.

Standard procedure: A written instruction that establishes what action is required, who is to act, and when the action is to take place.

Staphylococcus: Any of various spherical parasitic bacteria that occur in grapelike clusters and cause infections.

Strategic: Elements of incident management that are characterized by continuous, long-term, high-level planning by senior-level organizations.

Static pressure: The potential pressure exerted in all directions by a fluid at rest.

Status report: Information specifically related to the status of resources.

Steam sterilization: Treatment method for infectious waste using a saturated steam within a pressurized vessel such as an autoclave.

Sterilize: The use of a physical or chemical procedure to destroy all microbial life, including highly resistant spores.

Stochastic effect: Effect that occurs on a random basis independent of the size of dose of radiation.

Stoichiometry: Study of the mathematics of the material and energy balances (equilibrium) of chemical reactions.

Strategy: The general plan or direction selected to accomplish objectives.

Streptococcus: Any of various rounded, disease-causing bacteria that occur in pairs or chains.

Strontium (Sr): A silvery, soft metal that rapidly turns yellow in air. Sr-90 is one of the radioactive fission materials created within a nuclear reactor during its operation. Sr-90 emits beta particles during radioactive decay.

Substandard: A condition that deviates from what is acceptable, normal, or correct and is a potential hazard.

Surface burst: A nuclear-weapon explosion that is close enough to the ground for the radius of the fireball to vaporize surface material. Fallout from a surface burst contains very high levels of radioactivity.

Surge capability: The ability to manage patients requiring unusual or very specialized medical evaluation and care.

Surge capacity: The ability to evaluate and care for a markedly increased volume of patients—one that challenges or exceeds normal operating capacity.

Survey: Comprehensive study or assessment of a facility, workplace, or activity for insurance or loss-control purposes.

System: A clearly described functional structure, including defined processes, that coordinates otherwise diverse parts to achieve a common goal.

Systematic: Striving toward goal accomplishment in a planned manner using predetermined steps or procedures.

T

Tactical: ICS elements characterized by the execution of specific actions or plans in response to an actual incident.

Tactics: Deployment and directing of resources on an incident to accomplish the objectives designated by strategy.

Tendinitis: A condition where the muscle/tendon junction becomes inflamed.

Tenosynovitis: A condition that results in the inflammation of the tendons and their sheaths.

Teratogen: A substance or agent that, when a pregnant female is exposed to it, can cause malformations in the fetus.

Terrorism: Any premeditated, unlawful act dangerous to human life or public welfare that is intended to intimidate or coerce civilian populations or governments.

Thermoluminescent dosimeter: A badge that contains a thermoluminescent chip worn by persons working with or around radioactive materials.

Tingling: Pain or numbness in the thumb and first three fingers often characterize tingling; it is often associated with repeated wrist flexion.

TLV: An ACGIH-published TLV of an airborne concentration of a hazardous/toxic substance to which workers may be repeatedly exposed day after day without adverse effect.

Toxicity: Potential of a substance to have a harmful effect, and a description of the effect and the conditions or concentration under which the effect takes place.

Toxic substance: Any substance that can cause acute or chronic injury or illness to the human body.

Triage: The process of screening and classifying sick, wounded, or injured persons to determine priority needs to ensure the efficient use of resources.

Trigger finger: A condition caused by any finger being frequently flexed against some type of resistance.

U

Ultraviolet radiation: Electromagnetic radiation with wavelengths between soft X-rays and visible violet light.

Undetectable failure: A postulated failure mode in a FMEA for which there is no failure-detection method by which the operator is made aware of the failure.

Unified command: Agencies working together through their designated incident commanders or managers at a single location to establish a common set of objectives and strategies and a single incident action plan.

Uniform fire code: Regulations consistent with nationally recognized good practice for safeguarding life and property from the hazards of fire and explosion that arise from the storage, handling, and use of hazardous substances, materials, and devices.

Unity of command: Principle of management stating that each individual involved in incident operations will be assigned to only one supervisor.

Universal precautions: An OSHA terms for the method of infection control in which all human blood and certain other materials are treated as infectious for blood-borne pathogens.

Unstable: A chemical that, when in the pure state, will vigorously polymerize, decompose, condense, or become self-reactive under conditions of shock, pressure, or temperature.

Upper explosive limit: Highest concentration of a substance that will burn or explode when an ignition source is present, expressed in percent of vapor or gas in the air by volume.

Uranium: A naturally occurring radioactive element that is a hard, silvery-white, shiny metallic ore that contains a minute amount of uranium-234.

User seal check: An action conducted by the respirator user to determine if a respirator is properly seated to the face.

V

Vapor: The gaseous form of a substance that is normally in the solid or liquid state at room temperature and pressure.

Vapor density: The weight of a vapor or gas.

Vapor pressure: The pressure exerted by a saturated vapor above its own liquid in a closed container.

Vector: Organism that carries disease, such as insects or rodents.

Viscosity: The property of a liquid that causes it to resist flow or movement in response to external force applied to it.

Visible radiation (light): Electromagnetic radiation that can be detected by the human eye.

Volatile: The tendency or ability of a liquid to vaporize.

Volatile organic compounds: Compounds that evaporate from many housekeeping, maintenance, and building products made from organic chemicals.

W

Wavelength: The distance in the line of advance of a wave from any point to a like point on the next wave. Usually measured in angstroms, microns, micrometers, or nanometers.

Whole body count: The measure and analysis of the radiation being emitted from a person's entire body, detected by a counter external to the body.

Whole body exposure: An exposure of the body to radiation, in which the entire body, rather than an isolated part, is irradiated by an external source.

Wood alcohol: Methyl alcohol.

Work practice controls: Any controls that reduce the likelihood of exposure by altering the manner in which a task is performed.

X

Xenobiotic: A man-made substance, such as plastic, found in the environment.

X-ray: Electromagnetic radiation caused by deflection of electrons from their original paths, or inner orbital electrons that change their orbital levels around the atomic nucleus.

Suggested Reading

Abrahamson, E. *Change without Pain*. Boston, MA: Harvard Business School Press, 2004.

Adams, S.J. November 1997. Benchmarks of safety quality. *Professional Safety*, 33–34.

Agency for Toxic Substances and Disease Registry. Glossary. Accessed online, 2010. www.astdr.cdc.gov/glossary

Amar, A.D. June 1995. Principled versus analytical decision-making: Definitive optimization. *Mid-Atlantic Journal of Business*, 119.

American Industrial Hygiene Association. *Do I Work in a Sick Building?* New York, NY: ANSI, AIHA Essential Source, 1995.

American National Safety Standard ANSI/RIA R15.06-1992. *Industrial Robots and Robot Systems—Safety Requirements*. Ann Arbor, MI: Robotic Industries Association, 1992.

American National Standard Institute. *Manual on Uniform Traffic Control Devices for Streets and Highways*, ANSI D6.1. New York, NY: ANSI, 1971.

American National Standards Institute. *Standard Safety Levels With Respect to Human Exposure to Radiofrequency Fields*, ANSI C95.1-1991. New York, NY: ANSI, 1991.

American National Standards Institute. Standards Committee on the Safe Use of Lasers, ANSI Z136. New York, NY: ANSI, 1993.

American National Standards Institute. *Hazardous Industrial Chemicals—Precautionary Labeling*, ANSI Z-129.1-2006. New York, NY: ANSI, 2006.

American National Standards Institute. *Standard for the Provision of Slip Resistance on Walking/Working Surfaces*, ANSI/ASSE A1264.2. New York, NY: ANSI, 2006.

American National Standards Institute. *Safety Requirements for Workplace Walking Working Surfaces and Their Access: Floor, Wall and Roof Openings, Stairs and Guardrail Systems*, ANSI/ASSE A1264.1-2007. New York, NY: ANSI, 2007.

Anderson, R. *Security Engineering*. New York, NY: Wiley, 2001.

Andersson, R., and E. Lagerlöf. 1983. Accident data in the new Swedish information system on occupational injuries. *Ergonomics* 26: 33–42.

Apple, J. *Materials Handling Systems Design*. New York, NY: The Ronald Press, 1972.

Argyris, C., and D. Schon. *Organizational Learning II*. London: Addison-Wesley, 1996.

Arnold, H.J. Sanctions and rewards: Organizational perspectives. In *Sanctions and Rewards in the Legal System: A Multidisciplinary Approach*. Edited by M. L. Friedland. Toronto, ON: University of Toronto Press, 1989.

ASHE. *Electrical Standard Compendium*. Chicago, IL: American Society of Healthcare Engineering, 1999.

ASHRAE. *ASHRAE Handbook—HVAC Applications, I-P Version*. New York, NY: American Society of Heating, Refrigerating & Air-Conditioning Engineers, 2003.

ASSE. Return on Safety Investment White Paper. Des Plaines, IL: American Society of Safety Engineers, June 2002.

Athos, A.G., and R.C. Coffey. Time, space and things. In *Behavior in Organizations: A Multidimensional View*. Englewood Cliffs, NJ: Prentice Hall, 1975.

Bainbridge, L. 1983. Ironies of automation. *Automatica* 19:775–779.

Baker, S.P., B. O'Neil, M.J. Ginsburg, and G. Li. *Injury Fact Book*. New York, NY: Oxford University Press, 1992.

Bamber, L. 1979. Accident costing in industry. *Health and Safety at Work* (Croyden) 2/4:32–34.

Bare, A.R. 2006. A pressurization strategies for biocontainment. *R&D Laboratory Design Handbook* (11):59B63.

Barnard, C.I. *The Functions of the Executive*. Cambridge, MA: Harvard University Press, 1968.

Barnett, R., and D. Brickman. 1986. Safety hierarchy. *Journal of Safety Research* 17:49–55.

Barnett, T., and S. Valentine. 2004. Issue contingencies and marketers' recognition of ethical issues, ethical judgments, and behavioral intentions. *Journal of Business Research* 57:338–346.

Beauchamp, T., and N. Bowie. *Ethical Theory and Business*. Englewood Cliffs, NJ: Prentice Hall, 1993.

Behm, M., A. Veltri, and I. Kleinsorge. April 2004. Cost of safety: Cost analysis model helps build business case for safety. *Professional Safety* 49:22–29.

Berger, E.H., W.D. Ward, J.C. Morrill, and L.H. Royster (Editors). *Noise and Hearing Conservation Manual*, 4th Ed. Akron, OH: American Industrial Hygiene Association, 1986.

Berry, M. *Protecting the Built Environment: Cleaning For Health*. Chapel Hill, NC: TRICOM 21st Press, 1994.

Bird, F., and G. Germain. *Practical Loss Control Leadership*. Loganville, GA: International Loss Control Institute, 1985.

Birky, B., L. Slaback, and B. Schleien. *Handbook of Health Physics and Radiological Health*, 3rd Ed. Philadelphia, PA: Lippincott Williamson and Wilkins, 1998.

Blackwell, D.S., and G.S. Rajhans. *Practical Guide to Respiratory Usage in Industry*. Boston, MA: Blackwell Publishing, 1985.

Borisoff, D., and D.A. Victor. *Conflict Management: A Communication Skills Approach*. Englewood Cliffs, NJ: Prentice Hall, 1989.

Bothe, K.R. *World Class Quality*. New York, NY: AMACOM, 1991.

Bou, J.C., and I. Beltran. 2005. Total quality management, high-commitment human resource strategy and firm performance: An empirical study. *Total Quality Management* 16(1): 71–86.

Boulding, K.E. April 1956. General systems theory, the skeleton of science. *Management Science* 2: 197–208.

Brauer, R. *Safety and Health for Engineers*. New York, NY: Van Nostrand Reinhold, 1990.

Bronstein, D.A. *Demystifying the Law*. Chelsea, MI: Lewis Publishers, Inc., 1990.

Brookhuis, K., A. Hedge, H. Hendrick, E. Salas, and N. Stanton. *Handbook of Human Factors and Ergonomics Models*. Boca Raton, FL: CRC Press, 2005.

Bureau of Labor Statistics, U.S. Department of Labor. *Occupational Injuries and Illnesses in the United States by Industry*. Washington, DC: Bureau of Labor Statistics.

Burns, T., and G.M. Stalker. *The Management of Innovation*. London: Tavistock, 1961.

Carroll, S., and D. Gillen. 1980. Are the classical management functions useful in describing managerial work? *Academy of Management Review* 12(1):38–51.

Carson, H.T., and D.B. Cox. *Handbook on Hazardous Materials Management*, 4th Ed. Rockville, MD: Institute of Hazardous Materials Management, 1992.

Centers for Disease Control and Prevention (CDC) and National Institutes of Health (NIH). *Biosafety in Microbiological and Biomedical Laboratories* (BMBL), 5th Ed. 2007.

Chandler, A.D., Jr. *Strategy and Structure*. Cambridge, MA: MIT Press, 1962.

Chandler, P., and J. Sweller. 1991. Cognitive load theory and the format of instruction. *Cognition and Instruction* 8(4): 293–332.

Chen, A.Y.S., and J.L. Rodgers. 1995. Teaching the teachers TQM. *Management Accounting* 76: 42–46.

Chesbrough, H.W. *Open Innovation: The New Imperative for Creating and Profiting from Technology*. Boston, MA: Harvard Business School Press, 2003.

Chubb Group of Insurance Companies. The Rewards of Managing Risks: A Guide for Entrepreneurs and Managers. Warren, NJ: Chubb Group of Insurance Companies, 1997.

Ciriello, V.M., P.G. Dempsey, R.V. Maikala, and N.V. O'Brien, Revisited: Comparison of two techniques to establish maximum acceptable forces of dynamic pushing for male industrial workers. *International Journal of Industrial Ergonomics* 37(11–12):877–892.

Cooper, D. June 2002. Safety culture. *Professional Safety*: 30–36.

Corbett, J.M. 1988. Ergonomics in the development of human centered AMT. *Applied Ergonomics* 19:35–39.

Crainer, S. November 1988. 75 Greatest Management Decisions Ever Made. *Management Review*.

Creech, B. 1994. *The Five Pillars of TQM*. New York, NY: Truman Talley Books/Dutton.

Criscimagna, N. *Practical Application of Reliability Centered Maintenance Report*. Rome, NY: Reliability Analysis Center, 2001.

Daft, R.L. 2004. Theory Z: Opening the corporate door for participative management. *Academy of Management Executive* 18(4):117–122.

David, M., and J. Wilbur. *Fault Tree Analysis Application Guide, Report No. FTA*. Rome, NY: Reliability Analysis Center, 1990.

Davids, M. 1999. W. Edwards Deming (1900–1993) Quality Controller. *Journal of Business Strategy* 20(5):31–32.

Davies, J.C., and D.P. Manning. 1994. MAIM: The concept and construction of intelligent software. *Safety Science* 17:207–218.

Dell, G. 1999. Safe place versus safe person, a dichotomy, or is it? *Safety Science Monitor* 3, Article 14, Special Edition.

Della-Giustina, D. *Motor Fleet Safety and Security Management*. Boca Raton, FL: CRC Press, 2004.

Dellinger, S., and B. Deane. *Communicating Effectively—A Complete Guide for Better Managing*. Radnor, PA: Chilton Book Company, 1982.

Deming, W.E. *Out of the Crisis*. Cambridge, MA: MIT Press, 1986.

Department of Education, Office of Research. Choosing the Right Training Program, 1994.

Department of Trade and Industry. *Leisure Accident Surveillance System (LASS): Home and Leisure Accident Research 1986 Data, 11th Annual Report of the Home Accident Surveillance System*. London: Department of Trade and Industry, 1987.

Dessler, G. *Human Resource Management*, 10th Ed. Englewood Cliffs, NJ: Pearson/Prentice Hall, 2004.

Dipilla, S. *Slip and Fall Prevention: A Practical Handbook*, 2nd Ed. Boca Raton, FL: CRC Press, 2010.

Douglas, M., and A. Wildavsky. *Risk and Culture*. Berkeley, CA: University of California Press, 1983.

Drucker, P.F. *Management: Tasks, Responsibilities, Practices*. New York, NY: Harper & Row, 1974.

Drucker, P.F. May/June 1990. The Emerging Theory of Manufacturing. *Harvard Business Review*.

Drucker, P.F. June 1, 2004. What Makes an Effective Executive? *Harvard Business Review*.

Dumas, J., and M. Salzman. *Reviews of Human Factors and Ergonomics*. Human Factors and Ergonomics Society, 2006.

Earley, P.C., S. Ang, and J. Tan. *CQ: Developing Cultural Intelligence in the Workplace*. Stanford, CA: Stanford University Press, 2005.

Earnest, R.E. November 1997. Characteristics of proactive and reactive safety systems. *Professional Safety* 42:27.

Eijkemans, G. and M. Fingerhut. 2005. Proceedings from the economic evaluation of health and safety interventions at the company level conference. *Journal of Safety Research* 36(3):207–308.

Environmental Protection Agency. Asbestos Waste Management Guidance. Washington, DC: EPA, 1985.

Equal Employment Opportunity Commission. *Department of Justice, Technical Assistance Manual (Title I), Americans with Disabilities Act*, 1992.

Erickson, J. May 1997. The relationship between corporate culture and safety performance. *Professional Safety* 42:29–33.

Esmail, R., D. Banack, C. Cummings, et al. October 2006. Is your patient ready for transport? Developing an ICU patient transport decision scorecard. *Healthcare Quarterly*, 9 Spec:80–86.

Fayol, H. *General and Industrial Administration*. London: Sir Issac Pitman & Sons, 1949.

Federal Emergency Management Agency, United States Fire Administration, National Fire Academy. *Recognizing and Identifying Hazardous Materials*. Participant Manual, 2nd Ed. Washington, DC: Government Printing Office, 1985.

Federal Emergency Management Agency. *Guide for All-Hazard Emergency Operations Planning: State and Local Guide 101*. New York, NY: FEMA, September 1996.

Federal Emergency Management Agency Web Site, Fact Sheet, Terrorism, 1998.

Federal Emergency Management Agency. *Disaster Planning Guide for Business and Industry*. New York, NY: FEMA, 2004.

Feiner, F., D. Miller, and F. Walker. *Chart of the Nuclides*, 13th Ed. San Ramon, CA: General Electric Corporation, 1983.

Fenix, F. *Safety Leadership, Quick, Easy and Cheap.* Malvern, AR: EZ Up Inc, 1999.

Ferrell, O.C., J. Fraedrich, and L. Ferrell. *Business Ethics.* Boston, MA: Houghton Mifflin Company, 2002.

Ferry, T.S. *Modern Accident Investigation and Analysis*, 2nd Ed. New York, NY: John Wiley & Sons, 1988.

Feyer, A.M., and A.M. Williamson. 1991. An accident classification system for use in preventive strategies. *Scandinavian Journal of Work Environment Health* 17:302–311.

Fiedler, F.E. *A Theory of Leadership Effectiveness.* New York, NY: McGraw-Hill, 1967.

Flood, R.L., and E.R. Carson. *Dealing With Complexity: An Introduction to the Theory and Application of Systems Science*, 2nd Ed. New York, NY: Plenum Press, 1993.

Flood, R.L., and M.C. Jackson. *Creative Problem Solving: Total Systems Intervention.* Chichester, UK: Wiley, 1991.

FMC. *Product Safety Sign and Label System.* Santa Clara, CA: FMC Corporation, 1985.

Frederiksen, L.W. *Handbook of Organizational Behavior Management.* New York, NY: Wiley, 1982.

Friend, M. February 1997. Examine your safety philosophy. *Professional Safety* 42:34.

Friend, M.A., and L.R. Pagliari. May 2000. Establishing a safety culture: Getting started. *Professional Safety* 45:30.

Garvin, D.A. 1993. *Building a Learning Organization.* Harvard Business Review, 71.

Gebrewold, F., and F. Sigwart. August 1997. Performance objectives: Key to better safety instruction. *Professional Safety* 42:25–27.

Geller, E.S. *Working Safe: How to Help People Actively Care For Health and Safety*, 2nd Ed. Boca Raton, FL: CRC Press, 1996.

Geller, S. May 2000. Ten leadership qualities for a total safety culture: Management is not enough. *Professional Safety* 44:16–19.

Gido, J. and J. Clements. *Successful Project Management.* Cincinnati, OH: South-Western College Publishing, 1999.

Gielen, A.C. 1992. Health education and injury control: Integrating approaches. *Health Education Quarterly* 19(2):203–218.

Gielo-Perczak, K., W.S. Maynard, and A. Didomenico. Multidimensional aspects of slips, trips, and falls. In *Reviews of Human Factors and Ergonomics*, Vol 2. Edited by Robert Williges. Santa Monica, CA: 2006.

Goldenhar, L.M., and P.A. Schulte. 1999. Intervention research in occupational health and safety. *Journal of Occupational Medicine* 36(7):763–775.

Goodstein, L.P., H.B. Anderson, and S.E. Olsen. *Tasks, Errors and Mental Models.* London: Taylor & Francis, 1988.

Gordon, H.M. *A Management Approach to Hazard Control.* Bethesda, MD: Board of Certified Hazard Control Management, 1994.

Green, L.W., and M.W. Kreute. *Health Promotion Planning: An Educational and Environmental Approach.* Mountainview, CA: Mayfield Publishing Company, 1991.

Greenberg, M.D., P. Chalk, H.H. Willis, et al. *Maritime Terrorism: Risk and Liability.* Santa Monica, CA: RAND Corporation, 2006.

Gresov, C., and R. Drazin. April 1997. Equifinality: Functional equivalence in organizational design. *Academy of Management Review* 22:403–428.

Grimaldi, J., and R. Simonds. *Safety Management*, 5th Ed. Homewood, IL: Irwin, 1989.

Guastello, S.J. *The Comparative Effectiveness of Occupational Accident Reduction Programs.* Paper presented at the International Symposium Alcohol Related Accidents and Injuries. Yverdon-les-Bains, Switzerland, December 2–5, 1991.

Haddon, W.J. 1972. A logical framework for categorizing highway safety phenomena and activity. *Journal of Trauma* 12:193–207.

Hale, A.R., and A.I. Glendon. *Individual Behavior in the Face of Danger.* Amsterdam, The Netherlands: Elsevier, 1987.

Hale, A.R., and M. Hale. 1970. Accidents in perspective. *Occupational Psychology* 44:115–122.

Hammer, W. *Occupational Safety Management and Engineering*, 3rd Ed. Englewood Cliffs, NJ: Prentice Hall, 1985.

Hansen, D.J. *The Work Environment, Occupational Health Fundamentals.* Chelsea, MI: Lewis, 1991.

Hare, V.C. *System Analysis: A Diagnostic Approach.* New York, NY: Harcourt Brace World, 1967.

Harms-Ringdahl, L. *Safety Analysis, Principles and Practice in Occupational Safety*, Vol. 289. Amsterdam, The Netherlands: Elsevier, 1993.

Harrison, R. *Diagnosing Organizational Culture. Trainer's Manual.* San Fransico, CA: Jossey-Bass, 1993.

Harvard Business School Press. *Harvard Business Review on Knowledge Management.* 6th Ed. Boston, MA: Harvard Business School Press, 1998.

Hayes, A.W., Ed. *Principles and Methods of Toxicology.* New York, NY: Raven, 1989.

Hazards and Regulation of Cleaning Chemicals, Health Care without Harm, Health Facilities Management Magazine, October 2006. AHA, Chicago.

Health and Safety Executive. *Road Transport in Factories and Similar Workplaces, Guidance Note GS9(R).* London: HMSO, 1992.

Heinrich, H. *Industrial Accident Prevention, a Scientific Approach*, 4th Ed. New York: McGraw-Hill, 1959.

Hodgson, E., and P.E. Levi. *A Textbook of Modern Toxicology.* New York: Elsevier Science, 1987.

Hosmer, L. *Ethics of Management.* Homewood, IL: Irwin, 1991.

Hugentobler, M.K., B.A. Israel, and S.J. Schurman. 1992. An action research approach to workplace health: Intergrating methods. *Health Education Quarterly* 19(1):55–76.

Humbert, P. Top Ten Tools For Effective Listening. Accessed online, January 24, 2011. www.philiphumbert .com/Articles/Listening.htm

Huotari, M.L. *Trust in Knowledge Management and Systems in Organizations.* Hershey, PA: Idea Group Publishing, 2003.

Huang, Y. H., Chen, P., Chen, J. C., DeArmond, S., and Cingularov, K. 2007. Roles of safety climate and shift work on perceived injury risk: A multi-level analysis. *Accident Analysis and Prevention* 39(6):1088–1096.

Incidence Rates for Nonfatal Occupational Injuries and Illnesses Involving Days Away from Work per 10,000 Full-Time Workers by Industry and Selected Events or Exposures Leading to Injury or Illness, Bureau of Labor Statistics (BLS). Accessed online, January 2012.

International Ergonomics Association. What Is Ergonomics? Accessed online, December 2010.

International Organization for Standardization (ISO). 196, Symbols, Dimensions, and Layout for Safety Signs, ISO R557. Geneva, ISO.

Jeffrey E.L. April 2000. What do accidents truly cost? *Professional Safety* 45:38–42.

Jervis, S., and T. Collins. September 2001. Measuring safety's return on investment. *Professional Safety.*

Kast, F.E., and J.E. 1972. Rosenzweig. General systems theory: Applications for organizations and management. *Academy of Management Journal* 15(4): 451.

Kastango, E. "Brutal Facts" from Overview USP 797 and Control of Contamination Presentation, Compliance Tools and Aseptic Certification for USP 797, STAR Center Training, 2007.

Katz, D., and R.L. Kahn. *The Social Psychology of Organizations.* New York, NY: John Wiley & Sons, 1978.

Keller, J.J., and Associates.Winter 2000. Why you've been handed responsibility for safety? *The Compass, Management Practice Specialty News.*

Kidd, P. Skill-based automated manufacturing. In *Organization and Management of Advanced Manufacturing Systems.* Edited by W. Karwowski and G. Salvendy. New York, NY: Wiley, 1994.

Kitzes, W.F. April 1991. Safety management and the consumer product safety commission. *Professional Safety.*

Kjellén, U. 1984. The deviation concept in occupational accident control, Part I: Definition and classification and Part II: Data collection and assessment of significance. *Accident Analysis and Prevention* 16:289–323.

Kjellén, U., and J. Hovden. 1993. Reducing risks by deviation control—A retrospection into a research strategy. *Safety Science* 16:417–438.

Kjellén, U., and T.J. Larsson. 1981. Investigating accidents and reducing risks—A dynamic approach. *Journal of Occupational Accidents* 3:129–140.

Klaassen, C., M. Amdur, and J. Doull (Editors). *Casarett and Doull's Toxicology: The Basic Science of Poisons*, 3rd Ed. New York: Macmillan Publishing Company, 1986.

Kleiman, L.S. *Human Resource Management: A Managerial Tool for Competitive Advantage.* Cincinnati, OH: South-Western College Publishing, 2000.

Kline, P., and B. Saunders. *Ten Steps to a Learning Organization.* Great River Books, Salt Lake City, UY, 1997.

Knapp, M.L., and J.A Hall. *Nonverbal Communication in Human Interaction*. Belmont, CA: Wadsworth Publishing, 2001.

Kohn, L.T., J.M. Corrigan, and M.S. Donaldson (Eds.). *To Err Is Human: Building A Safer Health System*. Washington, DC: National Academic Press, 2000.

Koontz, H., and C. O'Donnell. *Principles of Management: An Analysis of Managerial Functions*. New York, NY: McGraw-Hill Book Co., 1955.

Kotter, J.P., and J.L. Heskett. *Corporate Culture and Performance*. New York, NY: Free Press, 1992.

Krause, T. *Behavior-Based Safety Process: Managing Involvement for an Injury-Free Culture*. New York, NY: Van Nostrand Reinhold, 1999.

Krause, T. March 2000. Motivating employees for safety success. *Professional Safety* 45:22.

Krause, T. *Leading With Safety*. Hoboken, NJ: Wiley, 2005.

Lado, A.A., and M.C. Wilson. 1994. Human resource systems and sustained competitive advantage: A competency-based perspective. *Academy of Management Review* 19(4):699–727.

LaRue, B., and R.R. Ivany. December 2004. Transform your culture. *Executive Excellence*: 14–15.

Laser Institute of America. *Laser Safety Guide*, 4th Ed. Cincinnati, OH: Laser Institute of America, 1977.

Lawrence, P., and J. Lorsch. *Organization and Environment*. Homewood, IL: Irwin, 1969.

LeFranc, F. February 2005. A dynamic culture can make a franchise system successful. *Franchising World*: 75–77.

Lehto, M.R., and J.M. Miller. *Warnings: Volume I: Fundamentals, Design, and Evaluation Methodologies*. Ann Arbor, MI: Fuller Technical Publications, 1986.

Leigh, P., S. Markowitz, M. Fahs, and P. Landrigan. *Costs of Occupational Injuries and Illnesses*. Ann Arbor, MI: University of Michigan Press, 2000.

Leonard, D., and W.C. Swap. *Deep Smarts: How to Cultivate and Transfer Enduring Business Wisdom*. Boston, MA: Harvard Business School Press, 2005.

Leplat, J. 1978. Accident analyses and work analyses. *Journal of Occupational Accidents* 1:331–340.

Liberty Mutual Insurance Company. Liberty Mutual Executive Survey of Workplace Safety. Liberty Mutual Insurance Company, 2001.

Liberty Mutual Insurance Company. Liberty Mutual Workplace Safety Index, Liberty Mutual Insurance Company, 2009.

Lofgren, D. *Dangerous Premises*. New York: New York State School of Industrial & Labor Relations, Cornell University, 1989.

Loo-Morrey, M., and R. Houlihan. Further Slip-Resistance Testing of Footwear for Use at Work, British Health and Safety Executive, Health and Safety Laboratory, Report Number HSL, 2007.

Lowrance, W. *Of Acceptable Risk*. Los Altos, CA: William Kaufmann, Inc., 1976.

MacKenzie, E.J., D.M. Steinwachs, and B.S. Shankar. 1989. Classifying severity of trauma based on hospital discharge diagnoses: Validation of an ICD-9CM to AIS-85 conversion table. *Medical Care* 27:412–422.

Malhotra, Y. 2005. Integrating knowledge management technologies in organizational business processes: Getting real time enterprises to deliver real business performance. *Journal of Knowledge Management* 9(1): 7–28.

Manning, D.P. *Industrial Accident-Type Classifications—A Study of the Theory and Practice of Accident Prevention Based on a Computer Analysis of Industrial Injury Records*. MD Thesis. Liverpool, UK: University of Liverpool, 1971.

Manuelle, F. July 1997. Principles for the practice of safety. *Professional Safety*: 27–31.

Mathis, T. 2005. Lean behavior-based safety: How the process is evolving to survive in today's economy. *Occupational Hazards*.

Mayo, E. *The Human Problems of Industrial Civilization*. New York, NY: Macmillan, 1933.

McAfee, R.B., and A.R. Winn. 1989. The use of incentives/feedback to enhance work place safety: A critique of the literature. *Journal of Safety Research* 20:7–19.

Mcsween, T.E. *The Values-Based Safety Process: Improving Your Safety Culture with a Behavioral Approach*. New York, NY: Van Nostrand Reinhold, 1995.

Meredith, J.R., and S.J. Mantel, Jr. *Project Management: A Managerial Approach*, 5th Ed. New York, NY: John Wiley & Sons, 2002.

Meyer, M.W. *Theory of Organizational Structure*. Indianapolis, IN: Bobbs-Merrill, 1977.

Miller, J.M., M.R. Lehto, and J.P. Frantz. *Warnings and Safety Instructions: Annotated and Indexed*. Ann Arbor, MI: Fuller Technical Publications, 1994.

Mintzberg, H. *The Nature of Managerial Work*. New York, NY: Harper & Row, 1973.

Mohr, D.L. and D. Clemmer. 1989. Evaluation of an occupational injury intervention in the petroleum industry. *Accident Analysis and Prevention* 21(3):263–271.

Mooney, J., and A. Reiley. *Onward Industry*. New York, NY: Harper & Row, 1931.

National Council on Radiation Protection and Measurements. *Radiation Protection for Medical and Allied Health Personnel*, NCRP Report No. 48. Bethesda, MD: National Council on Radiation Protection and Measurements, 1976.

National Council on Radiation Protection and Measurements. *A Practical Guide to the Determination of Human Exposure to Radiofrequency Fields*, NCRP Report No. 119. Bethesda, MD: National Council on Radiation Protection and Measurements, 1993.

National Highway Traffic Administration. *Economic Burden of Traffic Crashes on Employers: Costs by State and Industry and by Alcohol and Restraint Use*. Publication DOT HS 809 682, 2003.

National Institute for Occupational Safety and Health. *Guide to Safety in Confined Spaces*. NIOSH Publication 87–113, 1987, 2006.

National Institute for Occupational Safety and Health. *Preventing Work-Related Musculoskeletal Disorders in Sonography*. NIOSH Publication 2006–148.

National Institute for Occupational Safety and Health/Environmental Protection Agency. *Building Air Quality: A Guide for Building Owners and Facility Managers*. Washington, DC: NIOSH/EPA, December 1991.

National Restaurant Association. *Safety Self-Inspection Program for Foodservice Operators*, 1988.

National Safety Council. *Accident Prevention Manual for Business and Industry*, 10th Ed. Itasca, IL: National Safety Council, 1992.

National Safety Council. *Fundamentals of Industrial Hygiene*, 3rd Ed. Itasca, IL: National Safety Council, 1988.

National Safety Council. *Product Safety Management Guidelines*. Itasca, IL: National Safety Council, 1989.

National Safety Council. *Public Employee Safety and Health Management*. Itasca, IL: National Safety Council, 1990.

National Safety Council. *Supervisors Safety Manual*, 8th Ed. Itasca, IL: National Safety Council, 1993.

Noe, R.A., J.R. Hollenbeck, B. Gerhart, and P.M. Wright. *Human Resource Management: Gaining a Competitive Advantage*, 5th Ed. Boston, MA: McGraw-Hill, 2006.

Noji, E., and G. Kelen. *Manual of Toxicologic Emergencies*. Chicago, IL: Year Book Medical Publishers, 1989.

Noll, G., M. Hildebrand, and J. Yvorra. *Hazardous Materials: Managing the Incident*, 2nd Ed. Fire Protection Publications, Oklahoma State University, 1995.

Oden, H.W. *Managing Corporate Culture, Innovation, and Intrapreneurship*. Westport, CT: Quorum Books, 1997.

Organization for Economic Cooperation and Development. *Behavioral Adaptation to Changes in the Road Transport System*. Paris: OECD, 1990.

OSHA. *Chemical Hazard Communication*. OSHA Publication No. 3084, 1998.

OSHA. *General Industry Digest*. OSHA Publication No. 2201.

OSHA. *Construction Industry Digest*. OSHA Publication No. 2202.

OSHA. *Small Business Handbook*. OSHA Publication No. 2209.

OSHA. *Material Storing and Handling*. Publication No. 2236.

OSHA. *Training Requirements and Guidelines*. OSHA Publication No. 2254.

OSHA. *Job Hazard Analysis Guide*. OSHA Publication No. 3071, 2002.

OSHA. *Hearing Conservation*. OSHA Publication No. 3074.

OSHA. *Controlling Electrical Hazards*. OSHA Publication No. 3075.

OSHA. *Respiratory Protection*. OSHA Publication No. 3079.

OSHA. *Hand and Power Tools*. OSHA Publication No. 3080.

OSHA. *How to Plan For Workplace Emergencies and Evacuations*. OSHA Publication No. 3088.

OSHA. *Working Safely With Video Display Terminals*. OSHA Publication No. 3092.

OSHA. *Hazardous Waste and Emergency Response*. OSHA Publication No. 3114.

OSHA. *Control of Hazardous Energy Lockout-Tagout*. OSHA Publication No. 3120.

OSHA. *Principal Emergency Response and Preparedness Requirements*. OSHA Publication No. 3122.

OSHA. *Stairways and Ladders*. OSHA Publication No. 3124.

OSHA. *Process Safety Management Guidelines for Compliance*. OSHA Publication No. 3133.

OSHA. *Permit-Required Confined Spaces*. OSHA Publication No. 3138.

OSHA. *Industrial Hygiene*. OSHA Publication No. 3143.

OSHA. *Fall Protection in Construction*. OSHA Publication No. 3146.

OSHA. Scaffolding. OSHA Publication No. 3150.

OSHA. *Hazards Small Business Safety Management Series*. OSHA Publication No. 3157.

OSHA. *Recordkeeping*. OSHA Publication No. 3160.

OSHA. *Screening and Surveillance: A Guide to OSHA Standards*. OSHA Publication No. 3162.

OSHA. *Safety and Health Add Value*. OSHA Publication No. 3180.

OSHA. *Model Plans for Bloodborne Pathogens and Hazard Communications*. OSHA Publication No. 3186.

OSHA. *Preventing Mold-Related Problems in the Indoor Workplace*. OSHA Publication No. 3304.

OSHA. *Protecting Responders during Treatment/Transport of Victims of Hazardous Substance Releases*. OSHA Publication No. 3370.

OSHA. *Small Entity Compliance Guide for Respiratory Protection Standard*. OSHA Publication No. 9071.

Ouchi, W.G., and Z. Theory. *How American Business Can Meet the Japanese Challenge?* Reading, MA: Addison-Wesley Publishing, 1982.

Oxenrider, J. *Creative Root Cause Analysis*. Middlebury, IN: Center for Creative Teamwork, 1998.

Panico, C.R. December 2004. Culture's competitive advantage. *Global Cosmetic Industry* 172(12):58–60.

Pascale, R. *Managing on the Edge*. New York, NY: Simon & Schuster, 1990.

Peters, T.J., and R.H. Waterman, Jr. *In Search of Excellence: Lessons from America's Best Run Companies*. New York, NY: Harper & Row, 1982.

Petersen, D. October 1997. Accountability, culture and behavior. *Professional Safety*.

Peterson, D. *Analyzing Safety System Effectiveness*. New York, NY: Van Nostrand Reinhold, 1996.

Peterson, D. *Safety Management: A Human Approach*, 2nd Ed. Goshen, NY: Aloray, 1988.

Pope, W. *Managing For Performance Perfection*. Weaverville, NC: Bonnie Brae Publications, 1990.

Porter, E.H.. *Manpower Development: The System Training Concept*, 1st Ed. New York: Harper and Row, 1964.

Rasmussen, J., K. Duncan, and J. Leplat. *New Technology and Human Error*. Chichester, UK: Wiley, 1987.

Reason, J.T. *Human Error*. Cambridge: Cambridge University Press, 1990.

Reason, J.T. 2000. Human error: Models and management. *BMJ* 320: 768–770.

Reason, J.T. *Managing the Risks of Organizational Accidents*. Surrey, UK: Ashgate, 1997.

Reinhardt, P., and J. Gordon. *Infectious Waste Management*. Chelsea, MI: Lewis Publishers, 1991.

Robbins, S.P., and M. Coulter. *Management*. Upper Saddle River, NJ: Prentice Hall, 1999.

Robertson, L.S. *Injury Epidemiology*. New York, NY: Oxford University Press, 1992.

Robustelli, P., and J. Kullmann. *Implementing Six Sigma to Build Lasting Change*. White Paper. Princeton, NJ: Six Sigma Qualtec, 2006.

Rodrik, D. *Has Globalization Gone Too Far?* Washington, DC: Institute for International Economics, 1997.

Roger, B. *Safety and Health for Engineers*. New York, NY: Van Nostrand Reinhold, 1990.

Ross, J.E. *Total Quality Management*. Delray Beach, FL: St. Lucie Press, 1995.

Saari, J. 1992. Successful implementation of occupational health and safety programs in manufacturing for the 1990s. *Journal of Human Factors in Manufacturing* 2:55–66.

Salthammer, T. (Ed.) *Organic Indoor Air Pollutants: Occurrence, Measurement, Evaluation*. Germany: Wiley-VCH, 1999.

Schein, E.H. *Organizational Culture and Leadership: A Dynamic View*. San Francisco, CA: Jossey-Bass, 1995.

Schelp, L. 1988. The role of organizations in community participation—Prevention of accidental injuries in a rural Swedish municipality. *Social Science and Medicine* 26(11):1087–1093.

Schneid, T. *Americans with Disabilities Act: A Practical Guide for Managers*. New York, NY: Van Norstrand Reinhold, 1992.

Schneider, B. (Ed.). *Organizational Climate and Culture*. San Francisco, CA: Jossey-Bass, 1990.

Schwope, A., P. Costa, J. Jackson, and D. Weitzman. *Guidelines for Selection of Chemical Protective Clothing*, 3rd Ed. American Conference of Governmental Industrial Hygienists, 1987.

Scott, W R. *Organizations: Rational, Natural, and Open Systems*. Englewood Cliffs, NJ: Prentice Hall, 1981.

Senge, P. Art and practice of the learning organization, 1990. In *The New Paradigm in Business: Emerging Strategies for Leadership and Organizational Change*. Edited By Ray, M. and A. Rinzler. New York: World Business Academy, 1993.

Senge, P.M. *The Fifth Discipline: The Art and Practice of Learning Organization*. London: Century Business, 1993.

Shale, P. Top Ten Ways to Retain Good Employees. Healthy Workplace.Com.

Simon, H.A. *Administrative Behavior*. New York: Free Press, 1945.

Sinclair, T.C. *A Cost-Effectiveness Approach to Industrial Safety*. London: HMSO, 1972.

Smeltzer, L.R., J.L. Waltman, and D.L Leonard. *Managerial Communication: A Strategic Approach*. Needham, MA: Ginn Press, 1991.

Smith, G.S., and H. Falk. 1987. Unintentional injuries. *American Journal of Preventive Medicine* 5(Suppl):143–163.

Smith, G.S., and P.G. Barss. 1991. Unintentional injuries in developing countries: The epidemiology of a neglected problem. *Epidemiological Reviews*: 228–266.

Sotter, G. *Stop Slip and Fall Accidents*. Mission Viejo, CA: Sotter Engineering Company, 2000.

Spath P. L. *Error Reduction in Healthcare: A Systems Approach to Improving Patient Safety*. 2nd Ed. Chicago, IL: Health Forum, 2000.

Speir, R.O. August 1998. Punishment in accident investigation. *Professional Safety*, 29–31.

Spengler, J.D., J.M. Samet, and J.F. Mccarthy. *Indoor Air Quality Handbook*. McGraw–Hill, 2001.

Stanton, N., P. Salmon, G. Walker, et al. *Human Factors Methods: A Practical Guide for Engineering and Design*. Surrey, UK: Ashgate, 2005.

Steers, R.M., and L.W. Porter. *Motivation and Work Behavior*, 5th Ed. New York, NY: McGraw-Hill, 1991.

Sugimoto, N. Subjects and problems of robot safety technology. In *Occupational Safety and Health in Automation and Robotics*. Edited by K. Noro. London: Taylor & Francis, 1987.

Sulzer-Azaroff, B. 1982. Behavioral ecology and accident prevention. *Journal of Organizational Behavior Management* 2: 11–44.

Surry, J. *Industrial Accident Research: A Human Engineering Appraisal*. Ontario, Canada: University of Toronto, 1969.

Sweller, J. 1988. Cognitive load during problem solving: Effects on learning. *Cognitive Science* 12(1):257–285.

Taylor, F.W. *The Principles of Scientific Management*. New York, NY: Harper, 1917.

Thomas, M.D. November 1997. Reinforcing safety values in people. *Professional Safety*.

Tichy, N. *The Leadership Engine: Building Leaders at Every Level*. Dallas, TX: Pritchett & Associates, 1998.

Timm, P.R., and K.B. DeTienne. *Managerial Communication*. Englewood Cliffs, NJ: Prentice Hall, 1991.

Tweedy, J. *Healthcare Hazard Control and Safety Management*, 2nd Ed. Boca Raton, FL: CRC Press, 2005.

Uris, A. *101 of the Greatest Ideas in Management*. New York, NY: Wiley, 1986.

Veazie, M.A., D.D. Landen, T.R. Bender, and H.E. Amandus. 1994. Epidemiologic research on the etiology of injuries at work. *Annual Review of Public Health* 15:203–221.

Vinas, T. July 2002. Best Practices—Dupont: Safety Starts at the Top. Industryweek.Com.

Vincoli, J. *Basic Guide to System Safety*. New York, NY: John Wiley & Sons, 2006.

Von Bertalanffy, L. *General System Theory: Foundations, Developments, Applications*. New York, NY: Braziller, 1968.

Vouros, G.A. 2003. Technological issues towards knowledge-powered organizations. *Journal of Knowledge Management* 7(2):114–127.

Waganaar, W.A., P.T. Hudson, and J.T. Reason. 1990. Cognitive failures and accidents. *Applied Cognitive Psychology* 4:273–294.

Wallerstein, N., and R. Baker. 1994. Labor education programs in health and safety. *Occupational Medicine State of the Art Reviews* 9(2):305–320.

Walters, H. January 1998. Identifying and removing barriers to safe behaviors. *Professional Safety*.

Weeks, J.L. 1991. Occupational health and safety regulation in the coal mining industry: Public health at the workplace. *Annual Review of Public Health* 12:195–207.

Weick, K.E. *Sensemaking in Organizations*. Thousand Oaks, CA: Sage, 1995.

Westinghouse Electric Corporation. 1981. *Product Safety Label Handbook*. Trafford, PA: Westinghouse Printing Division.

Wickens, C., and J. Hollands. *Engineering Psychology and Human Performance*. Englewood Cliffs, NJ: Prentice Hall, 1999.

Wickens, C., J. Lee, Y. Liu, and B. Gorden. *An Introduction to Human Factors Engineering*, 2nd Ed. Englewood Cliffs, NJ: Prentice Hall, 1997.

Wilde, G.J. 1982. The theory of risk homeostasis: Implications for safety and health. *Risk Analysis* 2:209–225.

Williams, J. April 2002. Improving safety leadership. *Professional Safety*.

Williamson, A.M., and A.M. Feyer. 1990. Behavioral epidemiology as a tool for accident research. *Journal of Occupational Accidents* 12:207–222.

Winfrey, F.L., and J.L. Budd. Autumn 1997. Reframing strategic risk. *SAM Advanced Management Journal* 62(4):13–22.

Woodward, J. *Industrial Organization: Theory and Practice*. London: Oxford University Press, 1965.

Wren, D.A. *The Evolution of Management Thought*, 4th Ed. New York, NY: John Wiley & Sons, 1994.

Wright, G. January 2005. Realigning the culture. *Building Design and Construction* 46(1): 26–34.

Youngberg, B., and M. Hatlie (Editors). *Patient Safety Handbook*. Sudbury, MA: Jones and Bartlett Publishers, 2003.

Index

Printed in the United States
by Baker & Taylor Publisher Services